ANALYTICAL CHEMISTRY AND MICROCHEMISTRY

T0295225

SOLID-PHASE EXTRACTION

PROCEDURE, APPLICATIONS AND EFFECTS

ANALYTICAL CHEMISTRY AND MICROCHEMISTRY

Additional books and e-books in this series can be found
on Nova's website under the Series tab.

ANALYTICAL CHEMISTRY AND MICROCHEMISTRY

SOLID-PHASE EXTRACTION

PROCEDURE, APPLICATIONS AND EFFECTS

BEN BENSON
EDITOR

science publishers
New York

Copyright © 2019 by Nova Science Publishers, Inc.

We have partnered with Copyright Clearance Center to make it easy for you to obtain permissions to reuse content from this publication. Simply navigate to this publication's page on Nova's website and locate the "Get Permission" button below the title description. This button is linked directly to the title's permission page on copyright.com. Alternatively, you can visit copyright.com and search by title, ISBN, or ISSN.

For further questions about using the service on copyright.com, please contact:
Copyright Clearance Center
Phone: +1-(978) 750-8400   Fax: +1-(978) 750-4470   E-mail: info@copyright.com.

NOTICE TO THE READER

Library of Congress Cataloging-in-Publication Data

Names: Benson, Ben (Writer on chemistry), editor.
Title: Solid-phase extraction: procedure, applications, and effects / editor, Ben Benson.
Description: Hauppauge, New York: Nova Science Publishers, Inc., 2018. |
  Series: Analytical chemistry and microchemistry | Includes bibliographical references and index.
Identifiers: LCCN 2018047551 (print) | LCCN 2018048612 (ebook) | ISBN
  9781536145830 (ebook) | ISBN 9781536145823 (softcover)
Subjects: LCSH: Extraction (Chemistry) | Solid-phase analysis.
Classification: LCC QD63.E88 (ebook) | LCC QD63.E88 S63215 2018 (print) | DDC 543/.19--dc23
LC record available at https://lccn.loc.gov/2018047551

Published by Nova Science Publishers, Inc. † New York

CONTENTS

PREFACE

Conventional sample preparation techniques like solid phase extraction and liquid-liquid extraction prove to be tedious, time-consuming and oftentimes cumbersome. As a result, new sample techniques have been developed based on the two methods to optimize solvent consumption, miniaturize extraction and decrease environmental impact. In Solid-Phase Extraction: Procedure, Applications and Effects, the authors explore the recent developments and applications of solid phase microextraction techniques used in the preconcentration of personal care products.

Following this, this collection addresses the use of chemical compounds for agricultural purposes which can cause the contamination of the environment. These contaminants can remain in the plant and soil or migrate from soil to water. As such, the authors suggest that the detection and quantification of pesticide residues in waters is of great concern.

Membrane disk solid-phase extraction have been used to pretreat large-volume water samples because their large cross-section areas allow relatively high flow rates. Solid-phase extraction disks maintain their shape from the pretreatment process to the assaying step, making them useful for wet analyses as well as direct analyses.

The closing chapter examines the development of the used solid-phase extraction sorbents since the introduction of the technique until today. Sample preparation always plays a key role in the chemical analysis of a variety of matrices and is the most time-consuming and complicated step of the entire analytical procedure.

Chapter 1 - Personal care products (PCPs) are chemicals that are used as active ingredients in cosmetics, fragrances, shampoos, toothpaste and soaps. The PCPs are categorized as emerging pollutants, due to the lack of environmental regulations put in place on their accepted levels in the environment. Determination of these pollutants in various matrices usually requires highly efficient separation techniques because they are found in low concentrations. Conventional sample preparation techniques like solid phase extraction and liquid-liquid extraction prove to be tedious, time-consuming and oftentimes cumbersome. As a result, new sample techniques have been developed based on the two methods to optimize solvent consumption, miniaturize extraction and decrease environmental impact. Thus, the aim of this review is to explore the recent developments and applications of solid phase microextraction techniques used in the preconcentration of personal care products. These techniques include solid phase microextraction, dispersive microextraction, and their derivatives.

Chapter 2 - The use of chemical compounds for agriculture purposes can cause the contamination of the environment. These contaminants can remain in the plant and soil or migrate from soil to water and due to the water cycle, fluxes of them reach the aquatic environments. For this reason, the detection and quantification of pesticide residues in waters is of a great concern. Triazines are a group of herbicides widely worldwide used for control of weeds in many agricultural crops, as well as, railways roadside and golf courses. These compounds can be transformed through biotic and abiotic processes; therefore, the main degradation products should be included in current analytical methods. Triazines are considered an important class of chemical pollutants, hence included in the Endocrine Disruption Screening Program by the U.S. Environmental Protection Agency (2009). The European Union has also included simazine and atrazine in the list of 33 priority substances in the EU Water Framework Directive (2000/60/EC). Furthermore, the Directive 98/83/EC regulates the presence of pesticides in waters for human consumption and the Directive 2008/105/EC sets the Environmental Quality Standards for these compounds in surface water. The Directive 2013/39/EU, amending the Directives 2000/60/EC and 2008/105/EC, adds terbutryn to the list of priority substances and also establishes a maximum permitted concentration of 0.34, 2 and 4 μg/L for terbutryn, atrazine and simazine respectively. Considering the characteristics of these compounds, it is important to develop reliable, sensitive and fast analytical methods to determine the low-level herbicide residues. The determination of triazines and/or their degradation products in water samples comprises two steps: an extraction procedure and chromatographic analysis. Regarding extraction procedure, solid phase extraction (SPE) is the most commonly used extraction technique. Nonpolar SPE sorbents are generally selected for extracting triazines from water samples; however, the degradation products can be more efficiently extracted by using polar sorbents. In this way, different solid phases have been employed for triazines and their hydroxy and dealkylated products. Currently, SPE is being replaced by new analytical procedures that minimize the waste of organic solvents according with the principles of Green Chemistry. Thus, some micro-extraction methods have been applied for extraction of triazines in water as alternative to SPE. However, some of these techniques have drawbacks such as low sensitivity, poor recoveries and, in many cases, they are very laborious. In this chapter, the state-of-art about the determination of triazine herbicides in water samples using solid phase extraction and microextraction techniques along with a review in this subject is presented.

Chapter 3 - Solid-phase extraction (SPE) has been widely used in the pharmaceutical, industrial, and environmental fields to separate and/or preconcentrate inorganic and organic analytes. In particular, membrane disk SPEs have been used to pretreat large-volume water samples because their large cross-section areas allow relatively high flow rates. SPE disks maintain their shape from the pretreatment process to the assaying step, making them useful for wet analyses as well as direct analyses, e.g., X-ray fluorescence spectrometry and X-ray spectrometry. In addition, miniature SPE disks can be used in combination with portable analyzers for on-site analysis. In this chapter, rapid and simple methods combining disk SPE with several detection techniques for the determination of trace elements in water are described.

Chapter 4 - Sample preparation always plays a key role in the chemical analysis of a variety of matrices and is the most time-consuming and complicated step of the entire analytical procedure. Solid-phase extraction (SPE) has been one of the most widely applied innovative sample preparation techniques, and since its introduction as an alternative to the liquid-liquid extraction (LLE), it has been presenting numerous advantages. Its benefits, compared to other techniques include simplicity and rapidity, which are both significant for the modern analytical and green chemistry demands, as well as selectivity, good repeatability and recoveries, low limits of detection (LODs), use of low solvent amounts while no specialized equipment is required. The development of SPE has been quite rapid throughout the years and this book chapter examines the evolution of the used SPE sorbents since the introduction of the technique until nowadays.

In: Solid-Phase Extraction
Editor: Ben Benson

ISBN: 978-1-53614-582-3
© 2019 Nova Science Publishers, Inc.

Chapter 1

RECENT DEVELOPMENTS AND APPLICATION OF SOLID PHASE MICROEXTRACTION FOR DETERMINATION OF PERSONAL CARE PRODUCTS

*Geaneth P. Mashile, Anele Mpupa and Philiswa N. Nomngongo**

Department of Applied Chemistry, University of Johannesburg,
Doornfontein Campus, Johannesburg, South Africa

ABSTRACT

Personal care products (PCPs) are chemicals that are used as active ingredients in cosmetics, fragrances, shampoos, toothpaste and soaps. The PCPs are categorized as emerging pollutants, due to the lack of environmental regulations put in place on their accepted levels in the environment. Determination of these pollutants in various matrices usually requires highly efficient separation techniques because they are found in low concentrations. Conventional sample preparation techniques

* Corresponding Author Email: pnnomngongo@uj.ac.za or nomngongo@yahoo.com.

like solid phase extraction and liquid-liquid extraction prove to be tedious, time-consuming and oftentimes cumbersome. As a result, new sample techniques have been developed based on the two methods to optimize solvent consumption, miniaturize extraction and decrease environmental impact. Thus, the aim of this review is to explore the recent developments and applications of solid phase microextraction techniques used in the preconcentration of personal care products. These techniques include solid phase microextraction, dispersive microextraction, and their derivatives.

Keywords: solid phase microextraction, personal care products, dispersive solid phase microextraction, magnetic solid phase microextraction

INTRODUCTION

There is a great interest in the development of efficient analytical procedure which plays a major role in various scientific disciplines. These disciplines are largely connected by analytical chemistry which is a branch of chemistry that supports numerous scientific realms including forensics and biomedical sciences (Tagliaro et al., 1997), environmental engineering (Orella et al., 1998), food industry (Baggiani et al., 2007), pharmaceutical science (Zacharis et al., 2008), amongst others. The emphasis on analytical procedures is highly based on the critical step called sample preparation (namely, sample clean-up, separation, preconcentration and fractionation). The sample preparation step is said to utilise an average of up to 61% of the total time required in the whole analytical process (Safarikova and Safarik., 1999). This is a critical step as a great amount of target analyte can be lost and any mistakes in the collection and processing could result in substantial errors in the final outcome despite the excellent performance of the analytical technique to be used (Wierucka and Bizuik, 2014). Moreover, the establishment of clean-up and preconcentration steps is a vital requirement for the environmental protection, food legislation and health authorities ((Rodriquez- Mozaz et al., 2009; Pićo et al., 2007). Due to present of pollutants in food and environmental samples at low concentrations (ng/g or ng/L) and often times associated with highly complex matrices (Azzouz and Ballesteros, 2012), sample preparation step become the only solution.

Consequently, solid phase extraction (SPE) has been the most widely used procedure for clean-up, extraction, separation, fractionation and preconcentration of trace pollutants from environmental, biological, food and beverage matrices (Liška et al., 2003; Biggiani et al., 2007; Aldelhelm et al., 2008; Pedruozo et al., 2007; Carlson et al., 2013). Solid phase extraction procedures are based on the adsorption or chelation of the analytes on the adsorbent (solid phase material) without changing the analyte's concentration and identity (Andrade- Eiroa et al., 2016).

Over 20 years, SPE has been used as the preferred sample preparation method (compared to over traditional liquid-liquid extraction (LLE)) in numerous EPA methods for analysis of various organic polltants in drinking water and wastewater (Andrade- Eiroa et al., 2016). This is due to its attractive advantages such as the ability to extract a wide range of organic analytes which can either be non- polar to very polar analytes from different samples (Zhoe et al., 2012). In addition, the use of different adsorbents makes the SPE method more attractive and effective. The disadvantages of traditional SPE include catridge blockage by particulates/ colloidal matter, use relatively large volumes of organic solvents, multiple steps, tedious and sometimes costly (Safari et al., 2017; Asati et al., 2017). To overcome the drawbacks of traditional SPE method, solid phase microextraction (SPME) have been developed as an alternative sample preconcentration method for analysis of personal care products (Dimpe and Nomngongo, 2016). As a result, numerous studies have been carried out for development various SPME methods for preconcentration of personal care products.

Therefore, this review attempts to provide insight into the recent developments in solid phase microextraction with emphasis on its application in the preconcentration and extraction of personal care products. The advantages, challenges and limitation of the SPME technique and possible improvements with the introduction of the different types of solid phase based microextraction, are highlighted.

SOLID PHASE MICROEXTRACTION

In recent years, the development of sample treatment methods has been focused on miniaturized extraction procedures, which eliminates the use of large volumes of chemicals, sorbent and organic solvents (Sajid., 2017). The advantages of these miniaturized procedures include simplicity, rapidity, and easy automation compared to conventional SPE procedures which suffer from poor extraction efficiency and fractionation of phase (Sarafraz-Yazdi and Amiri., 2010). Therefore, miniaturized procedures like solid phase microextraction (SPME) (Pawliszyn., 1990), dispersive liquid-liquid microextraction (DLLME)(Rezaee et al., 2006) and liquid phase microextraction (LPME) (He and Lee., 1997), amongst others, have been developed. However, among other microextraction, the development of SPME procedure by Arthur and Pawlisyzn 1990 has led to the establishment of different forms miniaturized SPE procedures (Sajid, 2017).

Studies have previously reported on the use of SPME in two different modes, that is, direct immersion mode (DI-SPME) (Arthur and Pawliszyn., 1990) or headspace mode (HS-SPME) (Zhang and Pawliszyn., 1993). However, each mode has unique challenges or limitations, for example, DI-SPME is not suitable for complex matrices such as food; sludge, soil and sediments. While HS-SPME is reported to be unable to extract non-volatile substances. As a result, further approaches to developing better functioning modes have been reported by Pawliszyn and co-workers (1996) and Basheer et al., (2004). Despite, the above mentioned specific limitations, SPME technique as a whole also presents some drawbacks such as carryover between extractions, fiber degradation with increased application and high cost due to dedicated and expensive apparatus, amongst others (Nojavan and Yazdanpanah., 2017). Therefore, these have led to the development of micro solid phase (μ-SPE) techniques which is believed to have minimal disadvantages compared to SPME.

Table 1. Summary of application of SPME for PCPs

Analyte(s)	Matrix	Adsorbent	Analytical technique	Recoveries	LOD	Reference
Benzoates, parabens, BHA, BHT and triclosan	Fish	PDMS, PDMS/DVB, and CAR/PDMS	LC–HRMS	56.3–119.9	0.05-2.00 µg/Kg	Zhang et al,
	Water		GC-MS	53 -111	0.013-0.55	Alvarez-Rivera et al, 2015
Triclosan, parabens	Sewage and sludge	Divinylbenzene/Carboxen/ Polydimethylsiloxane	GC-MS	70 - 90	10–20 ng/L	López-Serna et al, 2018
Fragrance allergens	Personal care products	β-cyclodextrin/graphene oxide-coated fiber			0.050–0.150	Hou et al, 2015
UV filters	Water samples	DVB/CAR/PDMS, PA or PDMS/DVB	GC-MS/MS	79.9 - 106	0.045-8.2	Vila et al, 2017

Table 2. Summary of application for Stir bar sorptive (micro)extraction (SBS(M)E

Analytes	Matrix	Coating	% Recovery	LOD(µg L−1)	Refs
Triclosan	Commercial toothpaste, saliva, urban wastewater samples	Polydimethylsiloxane (PDMS)	79	0.1	Silva and Nogueira, 2008
Triclosan, triclocarban, benzophenone-3, parabens	Wastewater	Polyethyleneglycol (PEG) modified silicone	Up to 80	0.02-0.04	Gilart et al., 2014
Musks	Vegetables and soil	PDMS	74-126	0.01-1.1 ng/g	Aguire et al., 2014
Triclocarban	Groundwater and wastewater	PDMS	92-96	0.01	Klein et al. 2010
Triclosan, triclocarban, benzophenone-3, parabens	Wastewater	Polar monolithic	2-64%	0.15-0.50	Gilart et al., 2013

In micro solid phase the sorbent material used for extraction and preconcentration of analytes is either sealed in porous polypropylene membrane envelope or coated on a stir bar (Basheer et al., 2009) as seen in Table 1.

Stir Bar Sorptive (Micro)Extraction (SBS(M)E

Stir bar sportive extraction is based on the principles of solid phase microextraction (Baltussen et al., 1999), in contrast, it uses higher sample volumes which increases the extraction efficiency (Mitra, 2003, Roldan-Pijuan et al., 2014). The SBS(M)E technique involves the addition of an adsorbent coated magnetic stirrer bar into a sample containing a vial (Chisvert et al., 2018). Once the extraction is completed, the stir bar is manually removed from the sample and the concentrated analyte can be eluted with a suitable eluent and analyzed with analytical detection techniques (Silva and Nogueira, 2010). The SBSE is said to be environmentally friendly, robust and highly reproducible (David and Sandra, 2007). Due to attractive propertied of SBSE, a number of researchers has employed this method for extraction and preconcentration of polar and no-polar personal care products.

Pedrouzo et al., (2010) reported the application of stir-bar-sorptive extraction (SBSE) with liquid desorption (LD) for preconcentration of four UV filters (2,2-dihydroxy-4-methoxybenzophenone, benzophenone-3, octocrylene, and octyldimethyl-p-aminobenzoic acid) and two antimicrobial agents (triclocarban and triclosan). The analytes were quantified by ultra-high-performance liquid chromatography–electrospray ionization triple-quadrupole tandem mass spectrometry (UHPLC–(ESI)MS–MS). Under optimum conditions, the method displayed low detection limits, and acceptable precision. Finally, the method successfully applied for determination of the analytes in real environmental waters.

In another study, Kawaguchi et al., (2008) developed a sample preparation method based on SBSE in situ derivatization for the simultaneous preconcentration of benzophenone (BP) and its derivatives in water samples. The determination step was carried out using thermal desorption (TD)–gas chromatography–mass spectrometry (GC–MS). The detection limit ranging from 0.5–2 ng L^{-1} for the analytes are obtained. In addition, the method demonstrated relatively wide linearity with the correlation coefficients of 0.990 for all the analyte. The spiked recovery tests were carried out and the average recoveries ranged between 102 and 128% with the relative standard deviation (RSD) less than 15%. Other applications of SBSE have be documented in the literature (Table 1, Benedé et al., 2016; Wooding et al., 2017; Suazo et al, 2017; Benedé et al., 2017; Pintado-Herrera et al., 2014; Lakade et al., 2015).

Other Sorptive Extraction Technique

It has been reported that the main limitation of SBSE is that only one non-polar coating based on poly(dimethylsiloxane) (PDMS) are readily available in the market (Lakade et al., 2015). Although this coating has demonstrated excellent performance for the adsorption and extraction of non-polar analytes, it is still not suitable for polar analytes (especially personal care products). To overcome this limitation, two new SBSE polar coatings that is Acrylate Twisters (based on polyacrylate with proportion of polyethylene glycol) and EG/Silicone Twisters (based on polyethylene glycol modified with silicone) were brought to the market (Gilart et al., 2014; Lakade et al., 2015). Despite those efforts, there is a need for more effect sorptive methods with better extraction performances than those obtained with the commercially available SBSE devices. For this recently, researcher have developed alternative sorptive methods such as capsule phase microextraction (CPME) (Lakade et al. 201) and fabric phase sorptive extraction (FPSE) (Lakade et al. 2016; Lakade et al., 2015; Kumar et al. 2017; Montesdeoca-Esponda et al. 2015) for extraction of polar personal care products.

Micro Solid Phase Extraction

The micro-solid phase extraction with a porous membrane (μ-SPE) were developed in 2006 (Sajid., 2017). This system utilize sorbents with a very small particle size of about 3 μm, compared to those of traditional SPE 50- 60μm with the high surface area (Alexandrou et al., 2014). Their cartridges are handheld and require only microlitres of solvent to condition the sorbent and elute the samples. Moreover, their cartridges are also reusable for up to 100 times with clean samples and the systems are sealable in order to reduce the loss of volatile compounds due to the external environment. As a result μ-SPE procedure provides a great potential in a wide range of applications, more especially in the environmental field. Compared to conventional LLE and SPE, μ-SPE as a sample preparation technique is qualified as a highly useful preparation technique to counter the challenges faced with conventional techniques (Sànchez-Gonzàlez et al., 2015). However, with its clear advantages, there are currently very few studies evaluating the performance of μ-SPE (Alexandrou et al., 2015). In micro solid phase extraction, just like any other pretreatment procedures, sorbents are employed based on the type of target analytes to be extracted extraction and there characteristic features as reviewed by Sajid, (2017).

Sorbent Selection in μ-SPE

The adsorbents depend on the nature of target analyte, as such μ-SPE present the possibility to employ various sorbent types based on the stipulated characteristics, namely:

1) Large-surface area, which can be readily functionalized with a wide range of analyte-binding organic functional groups with fast and well-established strategies (Sajid, 2017).

2) Excellent solid-state flow properties for accurate and reproducible measurements. This is because better solid-state flow properties allow for ease in packing sorbents into membrane envelopes (Sajid, 2017).

3) Non-adhesive to the membrane material, to allow for ease in the movement towards the bottom of membrane envelopes (Lim et al., 2013). Using sticky sorbents after packing can interfere with heat sealing of the open bag edge of the membrane (Lim et al., 2013; Sajid, 2017).

4) Stable (insoluble) both in extraction and desorption medium.

Different studies have reported the use of μ-SPE with a variety of sorbent material for the extraction of a wide range of target analytes. Sajid and co-workers (2015), investigated the use of C_2 based adsorbent for the extraction of polar compounds estrogens from ovarian cysts fluids samples. The C_2 sorbents used were C_8, C_{18}, Carbograph, Haye- Sep A, Haye- Sep B (28) due to estrogen affinity to for C_2 sorbent (Sajid et al., 2015. Similarly, Haye- Sep A was used as a sorbent for the extraction of parabens from ovarian cysts. This sorbent demonstrated better extraction capability compared to the other C_2 sorbents, due to the high affinity of Haye-Sep A to parabens (Sajid et al., 2015). Recently, Mashile et al, (2018) developed the, a simple, rapid and effective in-syringe micro-solid phase extraction (MSPE) combined with high performance liquid chromatography and a photo diode array detector (HPLC PDA) for separation, preconcentration and determination of parabens from environmental water samples. Chitosan-coated activated carbon (CAC) was used as the sorbent in homemade in-syringe MSPE device. Under optimum condition, adsorbent proved to be more effective in extraction of the target analyse in complex matrices. However, not many reports on the use of the μ-SPE procedure on the extraction and determination of personal care products were available at the time of this review.

Dispersive Solid Phase Microextraction

Dispersive solid phase microextraction (DSPME) was introduced by Anastassiadcs and colleagues in 2003 (Anastassiades et al., 2003). The principle is based on the direct addition of an adsorbent to the sample in order to carry out the extraction process (Fontana et al., 2011). The adsorbent dispersion is achieved by means of a vortex, shaking or ultrasonic bath (Chu and Letcher., 2015). Thereafter, the adsorbent is separated from the liquid solution, transferred to a desorption solvent before analysis (Fontana et al., 2011). The dispersion process allows the adsorbent to equally interact with the analyte, thus achieving great capacity per amount while significantly reducing channelling (Llompart et al., 2013). The DSPME procedure offers many advantages as compared to conventional SPE procedure, such as enhanced recoveries, less solvent consumption, easy to use, cost-effective and short analysis time (Basheer et al., 2007). Moreover, sorbent materials applied in this procedure improve its performance due to the presence of high reaction sites and large surface area (Asfaram et al., 2015). The most common technique for dispersion includes vortex assisted an ultrasound assisted dispersive solid phase microextraction amongst others. The application of DSPME in determination and preconcentration of PCPs is summarised in Table 3.

In vortex assisted SPME, the sorbent is completely mixed with the water sample using a vortex, this is said to permit increased interaction between the sorbent and target analyte (Rocio-Bautista et al, 2015). The use of the ultrasonic bath as means of dispersion has been substantiated by the fact that it uses ultrasonic waves instead of physical agitation and thus is said to be more effective (Celano et al, 2014).

Magnetic Solid Phase Microextraction

Initial work on magnetic solid phase extraction was published by Safarikova and Safaric (1999), where they developed an SPE procedure based on magnetic sorbents.

Table 3. Summary for application of DSPME of PCPs

Analyte(s)	Matrix	Adsorbent	Analytical technique	Recoveries (%)	LOD (µg/L)	Reference
Parabens	Water samples	HKUST-1	HPLC-DAD	78-80.3	0.1	Rocío-Bautista et al, 2015
Parabens	Water and cosmetics	Aminopropyl - functionalized magnetite nanoparticles	GC- PID	10-62		Abbasghorbani et al, 2013
Benzyl benzoate, hexylcinnamal, linalool and methyl paraben	Personal care products	Florisil	GC–MS/MS		0.0004-0.025 µg/g	Celeiro et al, 2014
benzophenone, N,N-diethyl-3-methylbenzamide and trichlorocarbanilide	Water samples	MPC@Al2O3-SiO2	HPLC-DAD	98–107	0.066 0.096	Mpupa et al, 2018
Benzophenone preservatives, triclosan	Sludge	C18-bonded silica	LC-MS/MS	50-107	0.117–5.55 ng/g	Li et al, 2015

Their experiments entailed the use of reactive copper phthalocyanine dye which was attached to silanized magnetite (blue magnetite) and magnetic charcoal as magnetic sorbents for pre-concentration of Safranin O and crystal violet dyes. The basic principle of MSPE is the use of a magnetic sorbent which enables the dispersion of the sorbent in a large sample volume which is followed by sorbent retrieval conducted by an external magnetic field (Maya et al., 2017). Isolation and elution of magnetic adsorbent with the analyte adsorbed on the surface are done by addition of suitable solvents (Ibarra et al., 2015). The magnetic particles are available in a wide range of sizes from to nanoscale to microparticles, but the type of the material is usually from 1 to 100nm (Wierucka and Biziuk, 2014). The magnetic core is usually made up of iron, nickel, cobalt or any of their oxides with magnetite (Fe_3O_4) (Vasconcelos and Fernandes, 2017). Recent research on magnetic separation has led to the development of new materials, for instance, a magnetic material containing a magnetite core coated with silica (Fe_3O_4/SiO_2) or polymer ($Fe_3O_4/Polymer$) (Gao et al., 2011). The addition of silica provided a gap for the introduction of functional groups of interest which play a role when surface modifications are required (Gao et al., 2010).

In analytical chemistry, the first application of magnetic solids was for the extraction and isolation of magnetic bacteria in marine sediments. These magnetotactic microorganisms contain magnetosomes arranged on a chain like manner which is made up of nanometer-sized membrane-bound crystals of magnetic iron metals (Blakemore., 1975) However, previous work had been done on magnetic MnO_2 adsorbent application was reported in 1996 by Towler et al., for recovery of Ra, Pb and Pl from seawater (Towler et al., 1996). Thereafter Safarikova introduced magnetic solid phase extraction for the preconcentration of target analytes in large sample volumes (Safarikova and Safaric., 1999). The use of magnetic material has since been used in various fields to enhance the separation of chemical substances from different natures. MSPE technique applications have spread across to analyses different analytical matrices where compounds such as pesticides, phenolic compounds, plastic softeners and detergents herbicides, dyes and heavy metals (Ibarra et al., 2015). Moreover, this has found its application in biosciences field where there is the isolation of proteins, peptides, cells (Aguilar-Arteaga et al., 2010), purification of RNA, DNA from viruses, bacteria and biological fluids (Sarkar and Irudayaraj., 2008).

The widespread use of magnetic nanoparticles (MNPs) as sorbents (MNPs) is due to advantages, such as their speed, (Li et al., 2010), applicability in extraction of analytes from large sample volumes, coating with provided functionalities and selectivity towards target analytes (Andrade- Eiroa et al., 2016). In addition, MNPs are recyclable with proper after use rinsing, desorption of analytes absorbed by a simple procedure like sonication (Chen at al., 2011), ease in preparation, handling and less solvent consumption with increased extraction recoveries. However, disadvantages of MNPs are also evident which include and not limited to, degradation of stationary phase and thermally unstable components during desorption procedure were high temperatures are used (Hassan et al., 1991). Furthermore, drawbacks such as the majority of target analytes are not an ion or charged molecules and have no interaction with magnetic fields. Moreover interaction between magnetic fields and charged moving particles are applicable only under certain circumstances i.e., (V ≠ 0) and not parallel or antiparallel to the magnetic field created by the magnetic adsorbent (Purcell and Morin., 2013).

Despite the drawbacks, efforts by analysts working on magnetic solid phase for preconcentration of organic pollutants are providing more enhancements to the physical fundamentals of this procedure. This involves the development of magnetic solid phase microextraction procedures which are better than the conventional SPE (Andrade- Eiroa et al., 2016). Yao et al., (2017) successfully applied magnatic microspheres (M88) to a portable MSPE device for the on- site –preconcentration of 11 PPCPs (triclosan, mefenamic, chloroamphenico, ketoprofen, clofibric, indometacin, acetylsalicyclic acid, bisphenol A, phenylphenol, gemfibrozil, ibuprofen) based on their different polarity from large volumes of aquatic environment samples. The extraction efficiency of M88 was comparable with Oasis HLP and proved to be higher than that of C18 cartridges. Futhermore, the method was validated for the selected PPCPs under optimum conditions with LOD ranging between 0.7-9.4 ng/L with excellent recoveries an reproducibility compounds. Moliner-Martinez and colleagues 2014 reported on MSPE procedure that overcame the length of the sample preparation step. They developed an in-tube solid phase microextraction (Magnetic-IT-SPME) procedure using silica supported Fe_3O_4 magnetic nanoparticle as sorbent (Moliner-Martinez et al., 2011) for the preconcentration of organophosphorus compounds from wastewater (Moliner- Martinez et al., 2014). This procedure proved to be simple, had fast, with good extraction efficiencies and limit of detection. Other reports for MSPME are reported by, Lan et al., (2014) for the use of magnetic-MIP- SPME for preconcentration of estrogens from milk powder. The limit of detection from this procedure was around 1.5-5.5 $ng.g^{-1}$, with good reproducibility and RSD lower than 7.1%. However, not much literature was available on the magnetic solid microextraction for extraction and preconcentration of personal care products was available during this the time of this review.

CONCLUSION

Different types of solid phase microextraction methods have been reviewed. For the literature, it was evident the application of SPME method on the analysis of personal products is currently receiving more attention in analytical chemistry. It is thus no surprise that SMPE and its variants have been used for the extraction and preconcentration of personal care products. Studies has shown that the use of different sorbents and coating have provides promising results in the determination of personal care products, thus leading to better sensitivity and accuracy and low detection limits.

REFERENCES

Aguilar-Arteaga, K., Rodriguez, J. A., & Barrado, E. (2010). Magnetic solids in analytical chemistry: a review. *Analytica Chimica Acta*, *674*(2), 157-165.

Ahmadi, M., Madrakian, T., & Afkhami, A. (2016). Solid phase extraction of amoxicillin using dibenzo-18-crown-6 modified magnetic-multiwalled carbon nanotubes prior to its spectrophotometric determination. *Talanta*, *148*, 122-128.

Anastassiades, M., Lehotay, S. J., Štajnbaher, D., & Schenck, F. J. (2003). Fast and easy multi residue method employing acetonitrile extraction/ partitioning and "dispersive solid-phase extraction" for the determination of pesticide residues in produce. *Journal of AOAC international*, *86*(2), 412-431.

Arabi, M., Ghaedi, M., & Ostovan, A. (2017). Synthesis and application of in-situ molecularly imprinted silica monolithic in pipette-tip solid-phase microextraction for the separation and determination of gallic acid in orange juice samples. *Journal of Chromatography B*, *1048*, 102-110.

Arthur, C. L., & Pawliszyn, J. (1990). Solid phase microextraction with thermal desorption using fused silica optical fibers. *Analytical chemistry*, *62*(19), 2145-2148.

Asati, A., Satyanarayana, G. N. V., & Patel, D. K. (2017). Vortex-assisted surfactant-enhanced emulsification microextraction combined with LC–MS/MS for the determination of glucocorticoids in water with the aid of experimental design. *Analytical and Bioanalytical Chemistry*, *409*(11), 2905-2918.

Asfaram, A., Ghaedi, M., Goudarzi, A., & Soylak, M. (2015). Comparison between dispersive liquid–liquid microextraction and ultrasound-assisted nanoparticles-dispersive solid-phase microextraction combined with microvolume spectrophotometry method for the determination of Auramine-O in water samples. *RSC Advances*, *5*(49), 39084-39096.

Baltussen, E., Sandra, P., David, F., & Cramers, C. (1999). Stir bar sorptive extraction (SBSE), a novel extraction technique for aqueous samples: theory and principles. *Journal of Microcolumn Separations*, *11*(10), 737-747.

Basheer, C., Alnedhary, A. A., Rao, B. M., & Lee, H. K. (2009). Determination of carbamate pesticides using micro-solid-phase extraction combined with high-performance liquid chromatography. *Journal of Chromatography A*, *1216*(2), 211-216.

Basheer, C., Chong, H. G., Hii, T. M., & Lee, H. K. (2007). Application of porous membrane-protected micro-solid-phase extraction combined with HPLC for the analysis of acidic drugs in wastewater. *Analytical chemistry*, *79*(17), 6845-6850.

Basheer, C., Pavagadhi, S., Yu, H., Balasubramanian, R., & Lee, H. K. (2010). Determination of aldehydes in rainwater using micro-solid-phase extraction and high-performance liquid chromatography. *Journal of Chromatography A*, *1217*(41), 6366-6372.

Benedé, J. L., Chisvert, A., Giokas, D. L., & Salvador, A. (2016). Determination of ultraviolet filters in bathing waters by stir bar sorptive–dispersive microextraction coupled to thermal desorption–gas chromatography–mass spectrometry. *Talanta*, *147*, 246-252.

Blakemore, R. (1975). Magnetotactic bacteria. *Science*, *190*(4212), 377-379.

Boyacı, E., Rodriguez-Lafuente, A., Gorynski, K., Mirnaghi, F., Souza-Silva, E. A., Hein, D., & Pawliszyn, J. (2015). Sample preparation with solid phase microextraction and exhaustive extraction approaches: Comparison for challenging cases. *Analytica chimica acta*, *873*, 14-30.

Celano, R., Piccinelli, A. L., Campone, L., & Rastrelli, L. (2014). Ultra-preconcentration and determination of selected pharmaceutical and personal care products in different water matrices by solid-phase extraction combined with dispersive liquid–liquid microextraction prior to ultra high pressure liquid chromatography tandem mass spectrometry analysis. *Journal of Chromatography A*, *1355*, 26-35.

Chen, C., Zhang, X., Long, Z., Zhang, J., & Zheng, C. (2012). Molecularly imprinted dispersive solid-phase microextraction for determination of sulfamethazine by capillary electrophoresis. *Microchimica Acta*, *178*(3-4), 293-299.

Chisvert, A., Benedé, J. L., & Salvador, A. (2018). Current trends on the determination of organic UV filters in environmental water samples based on microextraction techniques–A review. *Analytica Chimica Acta*.

Chu, S., & Letcher, R. J. (2015). Determination of organophosphate flame retardants and plasticizers in lipid-rich matrices using dispersive solid-phase extraction as a sample cleanup step and ultra-high performance liquid chromatography with atmospheric pressure chemical ionization mass spectrometry. *Analytica chimica acta*, *885*, 183-190.

David, F., & Sandra, P. (2007). Stir bar sorptive extraction for trace analysis. *Journal of Chromatography A*, *1152*(1-2), 54-69.

Fei, T., Li, H., Ding, M., Ito, M., & Lin, J. M. (2011). Determination of parabens in cosmetic products by solid- phase microextraction of poly (ethylene glycol) diacrylate thin film on fibers and ultra high- speed liquid chromatography with diode array detector. *Journal of separation science*, *34*(13), 1599-1606.

Fernández-Amado, M., Prieto-Blanco, M. C., López-Mahía, P., Muniategui-Lorenzo, S., & Prada-Rodríguez, D. (2016). Strengths and weaknesses of in-tube solid-phase microextraction: A scoping review. *Analytica Chimica Acta*, *906*, 41-57.

Fontana, A. R., Camargo, A., Martinez, L. D., & Altamirano, J. C. (2011). Dispersive solid-phase extraction as a simplified clean-up technique for biological sample extracts. Determination of polybrominated diphenyl ethers by gas chromatography–tandem mass spectrometry. *Journal of Chromatography A*, *1218*(18), 2490-2496.

Gao, Q., Lin, C. Y., Luo, D., Suo, L. L., Chen, J. L., & Feng, Y. Q. (2011). Magnetic solid- phase extraction using magnetic hypercrosslinked polymer for rapid determination of illegal drugs in urine. *Journal of Separation Science*, 34(21), 3083-3091.

Gao, Q., Luo, D., Ding, J., & Feng, Y. Q. (2010). Rapid magnetic solid-phase extraction based on magnetite/silica/poly (methacrylic acid–co–ethylene glycol dimethacrylate) composite microspheres for the determination of sulfonamide in milk samples. *Journal of Chromatography A*, *1217*(35), 5602-5609.

Ghorbani, M., Chamsaz, M., Rounaghi, G. H., Aghamohammadhasani, M., Seyedin, O., & Lahoori, N. A. (2016). Development of a novel ultrasonic-assisted magnetic dispersive solid-phase microextraction method coupled with high performance liquid chromatography for determination of mirtazapine and its metabolites in human urine and water samples employing experimental design. *Analytical and Bioanalytical Chemistry*, *408*(27), 7719-7729.

He, Y., & Lee, H. K. (1997). Liquid-phase microextraction in a single drop of organic solvent by using a conventional microsyringe. *Analytical Chemistry*, *69*(22), 4634-4640.

Ibarra, I. S., Rodriguez, J. A., Galán-Vidal, C. A., Cepeda, A., & Miranda, J. M. (2015). Magnetic solid phase extraction applied to food analysis. *Journal of Chemistry*, *2015*.

Jannesar, R., Zare, F., Ghaedi, M., & Daneshfar, A. (2016). Dispersion of hydrophobic magnetic nanoparticles using ultarsonic-assisted in combination with coacervative microextraction for the simultaneous preconcentration and determination of tricyclic antidepressant drugs in biological fluids. *Ultrasonics Sonochemistry*, *32*, 380-386.

Kanimozhi, S., Basheer, C., Narasimhan, K., Liu, L., Koh, S., Xue, F., & Lee, H. K. (2011). Application of porous membrane protected micro-solid-phase-extraction combined with gas chromatography–mass spectrometry for the determination of estrogens in ovarian cyst fluid samples. *Analytica Chimica Acta*, *687*(1), 56-60.

Kawaguchi, M., Ito, R., Honda, H., Endo, N., Okanouchi, N., Saito, K., Seto, Y. & Nakazawa, H. (2008). Simultaneous analysis of benzophenone sunscreen compounds in water sample by stir bar sorptive extraction with in situ derivatization and thermal desorption–gas chromatography–mass spectrometry. *Journal of Chromatography A*, *1200*(2), 260-263.

Kumar, R., Malik, A. K., Kabir, A., & Furton, K. G. (2014). Efficient analysis of selected estrogens using fabric phase sorptive extraction and high performance liquid chromatography-fluorescence detection. *Journal of Chromatography A, 1359*, 16-25.

Lakade, S. S., Borrull, F., Furton, K. G., Kabir, A., Fontanals, N., & Marcé, R. M. (2015). Comparative study of different fabric phase sorptive extraction sorbents to determine emerging contaminants from environmental water using liquid chromatography–tandem mass spectrometry. *Talanta, 144*, 1342-1351.

Lakade, S. S., Borrull, F., Furton, K. G., Kabir, A., Fontanals, N., & Marcé, R. M. (2015). Comparative study of different fabric phase sorptive extraction sorbents to determine emerging contaminants from environmental water using liquid chromatography–tandem mass spectrometry. *Talanta, 144*, 1342-1351.

Lakade, S. S., Borrull, F., Furton, K. G., Kabir, A., Marcé, R. M., & Fontanals, N. (2016). Dynamic fabric phase sorptive extraction for a group of pharmaceuticals and personal care products from environmental waters. *Journal of Chromatography A, 1456*, 19-26.

Lakade, S. S., Borrull, F., Furton, K. G., Kabir, A., Marcé, R. M., & Fontanals, N. (2018). Novel capsule phase microextraction in combination with liquid chromatography-tandem mass spectrometry for determining personal care products in environmental water. *Analytical and Bioanalytical Chemistry, 410*(12), 2991-3001.

Lan, H., Gan, N., Pan, D., Hu, F., Li, T., Long, N., & Qiao, L. (2014). An automated solid-phase microextraction method based on magnetic molecularly imprinted polymer as fiber coating for detection of trace estrogens in milk powder. *Journal of Chromatography A, 1331*, 10-18.

Li, B., Zhang, T., Xu, Z., & Fang, H. H. P. (2009). Rapid analysis of 21 antibiotics of multiple classes in municipal wastewater using ultra performance liquid chromatography-tandem mass spectrometry. *Analytica Chimica Acta, 645*(1-2), 64-72.

Lim, T. H., Hu, L., Yang, C., He, C., & Lee, H. K. (2013). Membrane assisted micro-solid phase extraction of pharmaceuticals with amino and urea-grafted silica gel. *Journal of Chromatography A, 1316*, 8-14.

Liu, F. J., Liu, C. T., Li, W., & Tang, A. N. (2015). Dispersive solid-phase microextraction and capillary electrophoresis separation of food colorants in beverages using diamino moiety functionalized silica nanoparticles as both extractant and pseudostationary phase. *Talanta*, *132*, 366-372.

Liu, R. L., Zhang, Z. Q., Jing, W. H., Wang, L., Luo, Z. M., Chang, R. M., ... & Fu, Q. (2016). β-Cyclodextrin anchoring onto pericarpium granati-derived magnetic mesoporous carbon for selective capture of lopid in human serum and pharmaceutical wastewater samples. *Materials Science and Engineering: C*, *62*, 605-613.

Llompart, M., Celeiro, M., Lamas, J. P., Sanchez-Prado, L., Lores, M., & Garcia-Jares, C. (2013). Analysis of plasticizers and synthetic musks in cosmetic and personal care products by matrix solid-phase dispersion gas chromatography–mass spectrometry. *Journal of Chromatography A*, *1293*, 10-19.

Ma, J. B., Qiu, H. W., Rui, Q. H., Liao, Y. F., Chen, Y. M., Xu, J., ... & Zhao, Y. G. (2016). Fast determination of catecholamines in human plasma using carboxyl-functionalized magnetic-carbon nanotube molecularly imprinted polymer followed by liquid chromatography-tandem quadrupole mass spectrometry. *Journal of Chromatography A*, *1429*, 86-96.

Mashile, G. P., Mpupa, A., Nomngongo, P. N., &Samanidou, V. F. (2018). In-Syringe Micro Solid-Phase Extraction Method for the Separation and Preconcentration of Parabens in Environmental Water Samples. *Molecules*, *23*(6).

Maya, F., Cabello, C. P., Frizzarin, R. M., Estela, J. M., Turnes, G., & Cerdà, V. (2017). Magnetic solid-phase extraction using metal-organic frameworks (MOFs) and their derived carbons. *TrAC Trends in Analytical Chemistry*.

Mei, M., & Huang, X. (2017). Online analysis of five organic ultraviolet filters in environmental water samples using magnetism-enhanced monolith-based in-tube solid phase microextraction coupled with high-performance liquid chromatography. *Journal of Chromatography A*, *1525*, 1-9.

Mitra, S. (Ed.). (2004). *Sample preparation techniques in analytical chemistry* (Vol. 237). John Wiley & Sons.

Moliner-Martinez, Y., Vitta, Y., Prima-Garcia, H., González-Fuenzalida, R. A., Ribera, A., Campíns-Falcó, P., & Coronado, E. (2014). Silica supported Fe3O4 magnetic nanoparticles for magnetic solid-phase extraction and magnetic in-tube solid-phase microextraction: application to organophosphorous compounds. *Analytical and bioanalytical chemistry, 406*(8), 2211-2215.

Montesdeoca-Esponda, S., Sosa-Ferrera, Z., Kabir, A., Furton, K. G., & Santana-Rodríguez, J. J. (2015). Fabric phase sorptive extraction followed by UHPLC-MS/MS for the analysis of benzotriazole UV stabilizers in sewage samples. *Analytical and bioanalytical chemistry, 407*(26), 8137-8150.

Pintado-Herrera, M. G., González-Mazo, E., & Lara-Martín, P. A. (2014). Atmospheric pressure gas chromatography–time-of-flight-mass spectrometry (APGC–ToF-MS) for the determination of regulated and emerging contaminants in aqueous samples after stir bar sorptive extraction (SBSE). *Analytica Chimica Acta, 851*, 1-13.

Piri-Moghadam, H., Ahmadi, F., & Pawliszyn, J. (2016). A critical review of solid phase microextraction for analysis of water samples. *TrAC Trends in Analytical Chemistry, 85*, 133-143.

Purcell, E. M., & Morin, D. J. (2013). *Electricity and magnetism.* Cambridge University Press.

Ramos, L. (2012). Critical overview of selected contemporary sample preparation techniques. *Journal of Chromatography A, 1221*, 84-98.

Rastkari, N., & Ahmadkhaniha, R. (2013). Magnetic solid-phase extraction based on magnetic multi-walled carbon nanotubes for the determination of phthalate monoesters in urine samples. *Journal of chromatography A, 1286*, 22-28.

Rezaee, M., Assadi, Y., Hosseini, M. R. M., Aghaee, E., Ahmadi, F., & Berijani, S. (2006). Determination of organic compounds in water using dispersive liquid–liquid microextraction. *Journal of Chromatography A, 1116*(1-2), 1-9.

Rocío-Bautista, P., Martínez-Benito, C., Pino, V., Pasán, J., Ayala, J. H., Ruiz-Pérez, C., & Afonso, A. M. (2015). The metal–organic framework HKUST-1 as efficient sorbent in a vortex-assisted dispersive micro solid-phase extraction of parabens from environmental waters, cosmetic creams, and human urine. *Talanta*, *139*, 13-20.

Roldán-Pijuán, M., Lucena, R., Cárdenas, S., &Valcárcel, M. (2014). Micro-solid phase extraction based on oxidized single-walled carbon nanohorns immobilized on a stir borosilicate disk: application to the preconcentration of the endocrine disruptor benzophenone-3. *Microchemical Journal*, *115*, 87-94.

Safari, M., Yamini, Y., Masoomi, M. Y., Morsali, A., & Mani-Varnosfaderani, A. (2017). Magnetic metal-organic frameworks for the extraction of trace amounts of heavy metal ions prior to their determination by ICP-AES. *Microchimica Acta*, *184*(5), 1555-1564.

Šafaříková, M., & Šafařík, I. (1999). Magnetic solid-phase extraction. *Journal of Magnetism and Magnetic Materials*, *194*(1-3), 108-112.

Sajid, M. (2017). Porous membrane protected micro-solid-phase extraction: A review of features, advancements and applications. *Analytica chimica acta*, *965*, 36-53.

Sajid, M., Basheer, C., Narasimhan, K., Choolani, M., & Lee, H. K. (2015). Application of microwave-assisted micro-solid-phase extraction for determination of parabens in human ovarian cancer tissues. *Journal of Chromatography B*, *1000*, 192-198.

Sánchez-González, J., Tabernero, M. J., Bermejo, A. M., Bermejo-Barrera, P., & Moreda-Piñeiro, A. (2015). Porous membrane-protected molecularly imprinted polymer micro-solid-phase extraction for analysis of urinary cocaine and its metabolites using liquid chromatography–Tandem mass spectrometry. *Analytica chimica acta*, *898*, 50-59.

Sarafraz-Yazdi, A., & Amiri, A. (2010). Liquid-phase microextraction. *TrAC Trends in Analytical Chemistry*, *29*(1), 1-14.

Silva, A. R. M., & Nogueira, J. M. F. (2010). Stir-bar-sorptive extraction and liquid desorption combined with large-volume injection gas chromatography–mass spectrometry for ultra-trace analysis of musk compounds in environmental water matrices. *Analytical and bioanalytical chemistry*, *396*(5), 1853-1862.

Skorek, R., Turek, E., Zawisza, B., Marguí, E., Queralt, I., Stempin, M., & Sitko, R. (2012). Determination of selenium by X-ray fluorescence spectrometry using dispersive solid-phase microextraction with multiwalled carbon nanotubes as solid sorbent. *Journal of Analytical Atomic Spectrometry, 27*(10), 1688-1693.

Souza-Silva, E. A., Jiang, R., Rodriguez-Lafuente, A., Gionfriddo, E., & Pawliszyn, J. (2015). A critical review of the state of the art of solid-phase microextraction of complex matrices I. Environmental analysis. *TrAC Trends in Analytical Chemistry, 71*, 224-235.

Suazo, F., Vásquez, J., Retamal, M., Ascar, L., & Giordano, A. (2017). Pharmaceutical compounds determination in water samples: comparison between solid phase extraction and stir bar sorptive extraction. *Journal of the Chilean Chemical Society, 62*(3), 3597-3601.

Teo, H. L., Wong, L., Liu, Q., Teo, T. L., Lee, T. K., & Lee, H. K. (2016). Simple and accurate measurement of carbamazepine in surface water by use of porous membrane-protected micro-solid-phase extraction coupled with isotope dilution mass spectrometry. *Analytica Chimica Acta, 912*, 49-57.

Towler, P. H., Smith, J. D., & Dixon, D. R. (1996). Magnetic recovery of radium, lead and polonium from seawater samples after preconcentration on a magnetic adsorbent of manganese dioxide coated magnetite. *Analytica Chimica Acta, 328*(1), 53-59.

Tsai, W. H., Huang, T. C., Huang, J. J., Hsue, Y. H., & Chuang, H. Y. (2009). Dispersive solid-phase microextraction method for sample extraction in the analysis of four tetracyclines in water and milk samples by high-performance liquid chromatography with diode-array detection. *Journal of Chromatography A, 1216*(12), 2263-2269.

Vasconcelos, I., & Fernandes, C. (2017). Magnetic solid phase extraction for determination of drugs in biological matrices. *TrAC Trends in Analytical Chemistry.*

Vila, M., Celeiro, M., Lamas, J. P., Dagnac, T., Llompart, M., & Garcia-Jares, C. (2016). Determination of fourteen UV filters in bathing water by headspace solid-phase microextraction and gas chromatography-tandem mass spectrometry. *Anal Methods, 8*(39), 7069-7079.

Wang, Z., Zhao, X., Xu, X., Wu, L., Su, R., Zhao, Y., & Dong, D. (2013). An absorbing microwave micro-solid-phase extraction device used in non-polar solvent microwave-assisted extraction for the determination of organophosphorus pesticides. *Analytica chimica acta*, *760*, 60-68.

Wierucka, M., & Biziuk, M. (2014). Application of magnetic nanoparticles for magnetic solid-phase extraction in preparing biological, environmental and food samples. *TrAC Trends in Analytical Chemistry*, 59, 50-58.

Wooding, M., Rohwer, E. R., & Naudé, Y. (2017). Comparison of a disposable sorptive sampler with thermal desorption in a gas chromatographic inlet, or in a dedicated thermal desorber, to conventional stir bar sorptive extraction-thermal desorption for the determination of micropollutants in water. *Analytica chimica acta*, *984*, 107-115.

Xu, L., & Lee, H. K. (2008). Novel approach to microwave-assisted extraction and micro-solid-phase extraction from soil using graphite fibers as sorbent. *J. Chromatogr A*, *1192*(2), 203-207.

Yahaya, N., Mitome, T., Nishiyama, N., Sanagi, M. M., Ibrahim, W. A. W., & Nur, H. (2013). Rapid dispersive micro-solid phase extraction using mesoporous carbon COU-2 in the analysis of cloxacillin in water. *Journal of Pharmaceutical Innovation*, *8*(4), 240-246.

Yao, Z., Zhao, Q., Ma, Y., Wang, W., Zhou, Q., & Li, A. (2017). Magnetic microsphere-based portable solid phase extraction device for on-site pre-concentration of organics from large-volume water samples. *Scientific Reports*, *7*(1), 8069.

Zhang, Z., & Pawliszyn, J. (1993). Headspace solid-phase microextraction. *Analytical chemistry*, *65*(14), 1843-1852.

In: Solid-Phase Extraction
Editor: Ben Benson

ISBN: 978-1-53614-582-3
© 2019 Nova Science Publishers, Inc.

Chapter 2

APPLICATION OF SOLID PHASE EXTRACTION FOR DETERMINATION OF TRIAZINE HERBICIDES IN ENVIRONMENTAL WATER SAMPLES

Elisa Beceiro-González[*]
and María José González-Castro

Grupo Química Analítica Aplicada (QANAP), Instituto Universitario
de Medio Ambiente (IUMA), Centro de Investigaciones Científicas
Avanzadas (CICA), Departamento de Química, Facultade de Ciencias,
Universidade da Coruña, Campus de A Coruña, A Coruña, Spain

ABSTRACT

The use of chemical compounds for agriculture purposes can cause
the contamination of the environment. These contaminants can remain in
the plant and soil or migrate from soil to water and due to the water cycle,
fluxes of them reach the aquatic environments. For this reason, the

[*] Corresponding Author Email: elisa@udc.es.

detection and quantification of pesticide residues in waters is of a great concern.

Triazines are a group of herbicides widely worldwide used for control of weeds in many agricultural crops, as well as, railways roadside and golf courses. These compounds can be transformed through biotic and abiotic processes; therefore, the main degradation products should be included in current analytical methods. Triazines are considered an important class of chemical pollutants, hence included in the Endocrine Disruption Screening Program by the U.S. Environmental Protection Agency (2009). The European Union has also included simazine and atrazine in the list of 33 priority substances in the EU Water Framework Directive (2000/60/EC). Furthermore, the Directive 98/83/EC regulates the presence of pesticides in waters for human consumption and the Directive 2008/105/EC sets the Environmental Quality Standards for these compounds in surface water. The Directive 2013/39/EU, amending the Directives 2000/60/EC and 2008/105/EC, adds terbutryn to the list of priority substances and also establishes a maximum permitted concentration of 0.34, 2 and 4 µg/L for terbutryn, atrazine and simazine respectively. Considering the characteristics of these compounds, it is important to develop reliable, sensitive and fast analytical methods to determine the low-level herbicide residues.

The determination of triazines and/or their degradation products in water samples comprises two steps: an extraction procedure and chromatographic analysis. Regarding extraction procedure, solid phase extraction (SPE) is the most commonly used extraction technique. Nonpolar SPE sorbents are generally selected for extracting triazines from water samples; however, the degradation products can be more efficiently extracted by using polar sorbents. In this way, different solid phases have been employed for triazines and their hydroxy and dealkylated products.

Currently, SPE is being replaced by new analytical procedures that minimize the waste of organic solvents according with the principles of Green Chemistry. Thus, some micro-extraction methods have been applied for extraction of triazines in water as alternative to SPE. However, some of these techniques have drawbacks such as low sensitivity, poor recoveries and, in many cases, they are very laborious.

In this chapter, the state-of-art about the determination of triazine herbicides in water samples using solid phase extraction and microextraction techniques along with a review in this subject is presented.

Keywords: triazines and degradation products, solid phase extraction, waters

1. INTRODUCTION

Triazines are a group of herbicides that are present in the ten most used herbicide formulations in Europe. Triazines have been extensively used as herbicides to provide pre- and post-emergence of grasses, crops and many weeds in cereals (Wang et al. 2010), but they are also employed for non-agricultural purposes including soil sterilization and road maintenance (Papadopoulos et al. 2012). Their indiscriminate use has impacts not only in soils, fruits and vegetables, but also in water when these pollutants are washed away by rain. After their application, a large proportion remains in the environment (Li et al. 2010) and because of their poor sorption, high persistence and high water solubility, these compounds migrate from soil to water by leaching and runoff (Andreu and Picó 2004).

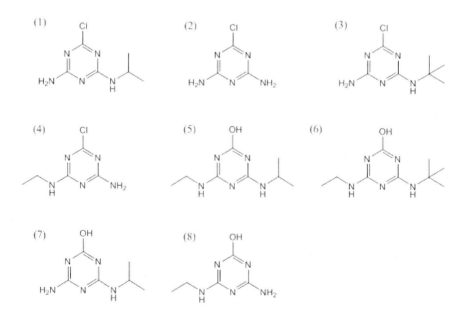

Figure 1. Structures of the eight main degradation products of triazines.
(1) DEA, (2) DEDIA, (3) DET, (4) DIA, (5) HA, (6) HT, (7) DEHA, (8) DIHA

The chemical contamination of the environment from herbicides is of great concern because they can remain many years in soil, water and organisms. Despite being persistent, they can be transformed in the environment through biotic and abiotic processes in the soil, groundwater and surface water (Jiang et al. 2005). In Figure 1, the main degradation products are presented: four dealkylated degradation products (desethyl-atrazine (DEA), desethyl-desisopropyl-atrazine (DEDIA), desethyl-terbuthylazine (DET) and desisopropyl-atrazine (DIA)), two hydroxylated degradation products (2-hydroxy-atrazine (HA) and 2-hydroxy-terbuthylazine (HT)) and two products resulting from the combination of dealkylation and hydroxylation (desethyl-2-hydroxy-atrazine (DEHA) and desisopropyl-2-hydroxy-atrazine (DIHA)). These degradation products can be even more toxic and persistent than the parent compounds. Because of their mobility in the soil-water environment, the degradation products can reach water bodies more easily than triazines; thus, the impact due to herbicides tends to be underestimated when only the triazines are analysed. Therefore, the main degradation products should be included in current analytical methods to obtain a better knowledge of water quality regarding herbicides pollution (Masiá et al. 2013a). In this way, studies examining the concentration of triazines in surface waters have expanded the list of compounds including their main degradation products (hydroxy and dealkylated products) (Bottoni et al. 2013, Köck-Schulmeyer et al. 2012).

During the last decade, the occurrence of micropollutants has been highlighted in thousands of publications, which have pointed out a growing concern about them. The chemical pollution of water can affect the environment and several effects as chronic toxicity on aquatics organisms, accumulation in the ecosystem, as well as injures in human health are well documented. Many of these substances, including triazine herbicides, are toxic and hazardous; for example, atrazine produces genotoxic damage in fish species (Santos and Martínez 2012). Therefore, triazine herbicides are considered as an important class of chemical contaminants and they are in the list of chemical pollutants that need to be more monitored. Atrazine and simazine have been included in the Endocrine Disruption Screening Program by the U.S. Environmental Protection Agency (Environmental

Protection Agency 2009). The European Union also included simazine and atrazine in the list of 33 priority substances in the EU Water Framework Directive (Commission Directive 2000/60/EC), by way of Decision 2455/2001/EC (Commission Decision 2455/2001/EC). Moreover, the Directive 2008/105/EC (Commission Directive 2008/105/EC) sets the Environmental Quality Standards (EQS) for these compounds in surface water and committees the Member States to set EQS for these compounds in sediments and biota at national level. Finally, the Directive 2013/39/EU includes terbutryn to the list of priority substances (Commission Directive 2013/39/EU). This Directive establishes a maximum permitted concentration of 0.34, 2 and 4 µg/L for terbutryn, atrazine and simazine respectively. Furthermore, Directive 2013/39/EU calls the attention on the important role of monitoring emerging pollutants which are not considered in monitoring programs but can have toxicological effects. On the other hand, the presence of pesticides in waters for human consumption is regulated by the European Directive 98/83/EC (Commission Directive 98/83/EC) that establishes the maximum admissible concentration of each pesticide at 0.1 µg/L and the total amount of pesticides at 0.5 µg/L.

Therefore, it is necessary to provide accurate, effective, simple and fast methods of analysis for these compounds that enable to improve the knowledge and data available on sources of these priority substances and ways in which pollution occurs. Furthermore, in order to support the implementation of the Directive 2013/39/EU, sensitive methods are required.

2. CHROMATOGRAPHIC ANALYSIS

The separation and identification of the triazine herbicides are accomplished using gas chromatography (GC) usually with mass spectrometry (MS) detection (Feizbakhsh and Ehteshami 2016; Fingler et al. 2017; Herrero-Hernández et al. 2017; Rimayi et al. 2018; Sanagi et al. 2015; Terzopoulou et al. 2015) or high-performance liquid chromate-graphy (HPLC) with different detectors such as ultraviolet detection (UV)

(Beale et al. 2010; Chen et al. 2014; González-Castro et al. 2016; Yang et al. 2015; Zhang et al. 2018), diode array detection (DAD) (Cao et al. 2017; Moliner-Martínez et al. 2015; Nasrollahpour and Moradi 2018; Papadopoulos et al. 2012; Rodríguez-González et al. 2015a; Serra-Mora et al. 2017; Zhou et al. 2017), or MS detection (Dujakovic et al. 2010; Huff and Foster 2011; Postigo et al. 2010; Rocha et al. 2015). Among these techniques, the application of liquid chromatography-tandem MS (LC-MS/MS) has provided an increased selectivity and sensitivity (Bolzan et al. 2015; De Toffoli et al. 2018; Rodríguez-González et al. 2017a; Roldán-Pijuán et al. 2015; Zhang et al. 2014; Zhou et al. 2018). Furthermore, the recent development of chromatographic columns using new stationary phases with particle size < 2 µm for ultra-pressure liquid chromatography (UPLC) allows significantly shorten the analysis time (Montoro et al. 2007). UPLC-MS/MS can offer not only good sensitivity but also high confidence in the confirmation of compounds detected allowing to achieve more than 3-4 identification points. Thus, UPLC coupled to tandem MS provides an interesting tool for fast determination of these compounds in water samples. UPLC-MS and UPLC-MS/MS have been used for determination of triazines in water samples (Barbosa et al. 2016; Cotton et al. 2016; Chen et al. 2015, Hurtado-Sánchez et al. 2013; Kalogridi et al. 2014; Loos et al. 2017; Rodríguez-González et al. 2016).

3. SOLID PHASE EXTRACTION (SPE)

During the last decade, a great progress has been done in methods of sample preparation for contaminants analysis; however, sample preparation remains to be one of the most critical stages in the analytical process. Trends for simplifying the analytical procedures have driven to the development of new analytical approaches which enable the determination of pollutants with improved capabilities, reduced clean-up and pre-concentration steps, and reduction of toxic reagents, solvent wastes and energy consumption, according to principles of green analytical chemistry.

Because of the complexity of the environmental samples and the relatively low herbicides concentration, analytes preconcentration and sample clean-up before the chromatographic analysis are mandatory to avoid interferences. Nevertheless, the main problem in pesticides extraction procedures is to obtain an efficient extraction of the analytes from matrix taking into account the low-levels in the sample.

Liquid-liquid extraction (LLE) is the classical method used to extract triazines and their degradation products from aqueous samples (Retamal et al. 2013). However, this technique requires large sample, high consumption of organic solvents, long extraction time, as well as a cleaning step to eliminate possible interferences. Nowadays, solid phase extraction (SPE) is widely accepted as an alternative extraction method to LLE for determination of contaminants in aqueous samples. SPE is the most commonly technique used for the preconcentration of triazines and their degradation products in waters because it offers considerable advantages such as: short time, low organic solvent consumption, cost effective, good enrichment factors and high recoveries. There is an extensive variety of commercially SPE sorbents available in cartridges and disks. Nonpolar SPE sorbents are generally selected for extracting triazines from water samples. By contrast, degradation products, which contain polar functional groups, require polar sorbents for their extraction (Benvenuto et al. 2010, Papadopoulos et al. 2007).

3.1. Hydrophilic Polymeric Sorbents

The most well known and widely used hydrophilic sorbent is Oasis HLB, which is a macroporous poly (N-vinylpyrrolidone-divinylbenzene) (PVP-DVB) copolymer. Among the advantages of this sorbent are the following: extraction of both polar and apolar compounds, cleaning of complex matrices and high capacity and efficiency to eliminate possible interferences (Gilart et al. 2014).

This sorbent has been employed in many studies for extraction and determination of triazines. Thus, Carabias et al. (2000) investigated two types of sorbents, C18 and Oasis HLB, for the determination of 15 herbicides, including atrazine and terbutryn, by HPLC-DAD. Oasis HLB was chosen because the cartridge-drying step, prior to the elution with ethyl acetate, was less critical and good recoveries were obtained. The limits of detection (LODs) obtained were 4 and 6 ng/L for atrazine and terbutryn respectively. The method was applied to the analysis of surface and ground water samples in agricultural areas of Zamora and Salamanca (Spain) (Carabias et al. 2003).

Dujakovic et al. (2010) have also compared C18 and Oasis HLB for the extraction of 14 pesticides belong to seven chemical classes (organophosphates, neonicotinoids, carbamates, diacylhydrazines, benzimidazoles, triazines and phenylureas) in water samples by LC-MS/MS. The triazines chosen for the study were atrazine, propazine and simazine. Regarding sorbents, the best recoveries for all tested pesticides were achieved using Oasis HLB cartridge due to its hydrophilic and lipophilic characteristics. The method has shown adequate recoveries (82-129%) and low limits of quantification (LOQs) for the triazines studied in surface and ground waters (1.9-13.1 ng/L).

A SPE procedure using Oasis HLB cartridges was employed for the determination of 70 organic micropollutants from various chemical classes (including ametryn, atraton, atrazine, prometryn, propazine, simazine, simetryn, terbuthylazine and terbutryn) in surface waters by GC-MS/MS (Terzopoulou et al. 2015). The recoveries of the triazines ranged from 86 to 106% with relative standard deviations (RSDs) from 0.84 to 23.8%. The LODs varied from 0.026 to 0.106 µg/L.

Other authors have used Oasis HLB for the simultaneous extraction of triazines and some of their metabolites in waters. Thus, a multiresidue method for the analysis of more than 40 herbicides (including atrazine, terbuthylazine, simazine, metribuzine, terbumeton, propazine, prometryn, sebuthylazine, DIA, DEA and DET) in waters has been developed by Montoro et al. (2007). The determination was carried out by UPLC-MS/MS. The LOQs ranged from 0.005 to 0.033 µg/L, which were below the values specified by the European Union. The recoveries obtained for water samples spiked at two levels (0.06 and 0.10 μg/L) were adequate (70-92% and 81-104% for 0.06 and 0.10 μg/L respectively).

Gervais et al. (2008) have also developed a multiresidue method employing this sorbent for the extraction of pesticides in water followed by UPLC-MS/MS. Among the pesticides under study, atrazine, cyanazine, DEA, DIA, simazine and terbuthylazine were selected. Five different cartridges (Envi Chrom P, Strata-X, M-N C18 Hydra, M-N easy and Oasis HLB) were assayed in order to achieve the suitable extraction conditions. Envi Chrom P showed the lowest recoveries whereas Oasis HLB exhibited the best results. The average recoveries of the triazines ranged from 92 to 107%. The LOQs of the method were satisfactory being the triazines correctly quantified at a concentration level of 0.03 μg/L (LOQs between 0.011 and 0.027 μg/L).

Hildebrandt et al. (2008) determined atrazine, DEA, DET, DIA, simazine and terbuthylazine in water samples using Oasis HLB cartridges followed by GC-MS analysis. The recoveries obtained were satisfactory (83-95%). Subsequently, Navarro et al. (2010) have used this procedure to determine 30 pesticides (including atrazine, DEA, propazine, simazine, terbuthylazine and terbutryn) in surface waters.

Oasis HLB cartridges have also been employed for the simultaneous determination of 30 compounds (including ametryn, atrazine, cyanazine, DEA, DET, DIA, prometryn, propazine, simazine, terbuthylazine and terbutryn) in waters (Mazzella et al. 2009). The quantifications were performed using LC-MS. The LOQs were from 0.001 to 0.032 μg/L for the triazines studied. In addition, the recoveries obtained for river water samples spiked at two different concentration levels (0.05 and 0.5 μg/L) showed adequate results, except for DEA and cyanazine at the lowest spiked level.

This sorbent has been also used to analyze 34 different polar organic contaminants (including atrazine, DEA, DET, simazine and terbuthylazine) in river water samples by LC-MS/MS (Loos et al. 2010).

Atrazine, simazine, terbuthylazine, terbumeton, terbutryn and their main transformation products (DEA, DET, desethylterbumeton (DETer), DIA, HA and HT) were determined in surface and wastewater samples using a pre-concentration step based on SPE followed by UPLC-MS/MS (Benvenuto et al. 2010). Two sorbents were tested (Oasis HLB and Oasis MCX). Oasis HLB was selected due to its ability to retain both non-polar and polar compounds affording much better recoveries (70-120%) and RSDs for all compounds.

Huff and Foster (2011) have developed a method for the determination of herbicides including s-triazines (atrazine, propazine and simazine) and their degradation products (DEA, DEDIA, DEHA, DIA, DIHA and HA) from surface waters. Prior to LC-MS determination, the herbicides were extracted using Oasis HLB cartridges. Spiked filtered river water yielded SPE recoveries ranging from 94.2 ± 4.8% for s-triazines and degradation products except for three compounds (DEDIA, DEHA and DIHA), with recoveries < 65%.

An analytical method based on SPE using Oasis HLB followed by GC-MS was also developed for the quantification of 39 pesticides including DEA, simazine, simetryn and terbuthylazine in estuarine water samples (Rocha et al. 2012). The compounds showed adequate recoveries (between 76.7 and 112.8%) and the method gave good LODs (0.06, 0.14, 0.17 and 0.2 µg/L for simazine, terbuthylazine, simetryn and DEA respectively).

Herrero-Hernández et al. (2012) compared four different sorbents: Oasis HLB, LiChrolut EN, C18 and graphitized carbon for the extraction of pesticides (herbicides, fungicides, insecticides and degradation products) from ground water. The analysis was performed by GC-MS. Five triazines (atrazine, propazine, simazine, terbuthylazine and terbutryn) and three degradation products (DEA, DET and DIA) were determined. The highest recoveries (65-108%) in the SPE procedure were achieved when 500 mL of water was preconcentrated using an Oasis HLB cartridge, and acetonitrile and acetone as elution solvents. The LODs were 4-26 ng/L for triazines and their degradation products. The authors applied this method to study the seasonal distribution of herbicide and insecticide residues in the water resources (surface and ground water) from a vineyard region of La Rioja (Spain) (Herrero-Hernández et al. 2017). The herbicide terbuthylazine and its metabolite DET were the compounds more frequently detected.

The combination of LC-MS/MS and UPLC-QTOF-MS has been applied to the analysis of surface and wastewater samples after conventional SPE using Oasis HLB cartridges (Masiá et al. 2013b). Among the compounds studied, atrazine, propazine, simazine, terbutryn, DEA and DIA were selected. The LOQs for triazines ranged from 1 to 6 ng/L and recoveries were higher than 70% for all analytes, except for omethoate and atrazine-deisopropyl, which yielded recoveries of 48% and 52%, respectively. The water extracts, previously analysed by LC-MS/MS, were also analysed by UPLC-QTOF-MS allowing the detection of more target compounds because of its higher sensitivity.

Oasis HLB has been also applied for the determination of 253 multiclass pesticides including triazines (atrazine, DEA, sebuthylazine, simazine and terbuthylazine) in lakes waters from Greece by UPLC-MS (Kalogridi et al. 2014).

Barbosa et al. (2016) optimized an SPE-UPLC-MS/MS method for the analysis of seven pesticides (including atrazine and simazine) and twelve pharmaceuticals in tap, fountain and well water. In order to assess the best performance of SPE cartridges, they compared Oasis HLB, MCX and MAX. Oasis HLB was the sorbent selected because of its versatility and suitability for most of compounds. The recoveries were 84.9% for simazine and 92.3% for atrazine, and the LODs were 0.15 and 0.12 ng/L for simazine and atrazine respectively.

Loos et al. (2017) carried out a study of emerging contaminants, including terbutryn, in water and other matrices in the Joint Danube Survey. The water samples (1 L) were extracted by automatic SPE with Oasis HLB (200 mg) cartridge employing 10 mL of ethyl acetate for elution. The analysis was performed by UPLC-MS/MS. The LOQ for terbutryn was 0.64 ng/L.

It is important to note that Oasis HLB has shown to have better ability to retain both triazines and some degradation products (DEA, DET, DETer, DIA, HA and HT) than other sorbents (Benvenuto et al. 2010, Gervais et al. 2008, Herrero-Hernández et al. 2012).

Others hydrophilic polymeric sorbents have been used for SPE of triazines. Thus, Matamoros et al. (2010) have used Strata-X as sorbent to develop an analytical procedure based on comprehensive two dimensional gas chromatography (GC × GC) coupled with time-of-flight mass spectrometry (TOF-MS) for the simultaneous determination of 97 organic contaminants at trace concentration in river water. The target triazines were ametryn, atrazine, prometon, prometryn, propazine, simazine, terbuthylazine and terbutryn. The best resolution of the target analytes in the contour plots was obtained when a nonpolar stationary phase was used in the first dimension and a polar one in the second dimension. The LOQs for triazines ranged from 5 to 26 ng/L and good recoveries (between 88 and 102%) were achieved.

Rimayi et al. (2018) studied the degree of triazine pollution in the Hartbeespoort Dam (South Africa) and their seasonal variation in lake, river and groundwater. SPE with Bond Elut Plexa (200 mg styrene divinyl benzyl) and 1 L sample were employed. Elution with 6 mL of dichloromethane and analysis by GC-MS was carried out for ametryn, atrazine, gesatamin prometon, simazine and terbuthylazine. Elution with 3 mL of dichloromethane and 3 mL of acetonitrile and analysis by LC-MS/MS was employed for DEA, DEDIA, DET, DIA, DIHA, HA, HT, prometryn and propazine. The recoveries for triazines ranged from 80 to 102% and 69 to 112% for degradation products. The LODs obtained were between 0.1 and 5 ng/L for triazines and between 0.1 and 0.2 ng/L for degradation products.

3.2. Hydrophobic Sorbents

The most used apolar sorbent for the extraction of triazines in water is octadecylsilane (C18) because of its ability to retain both apolar and lightly polar analytes. Thus, different studies using C18 as sorbent have shown acceptable recoveries for these compounds.

SPE using C18 cartridge connected to GC with TOF-MS has been used to a multiclass screening method of organic contaminants (including atrazine, DEA, DET, DIA, simazine, terbumeton and terbuthylazine) in natural and wastewater (Portolés et al. 2011). Surface water, ground water and effluent wastewater were spiked with all target analytes at three concentration levels (0.02, 0.1 and 1 µg/L). The wide majority of compounds investigated were correctly identified in the samples spiked at 1 µg/L. The identification was more problematic when analyte concentration was less than 0.1 µg/L, especially in complex-matrix samples. However, many contaminants could be properly identified at the lowest level 0.02 µg/L in cleaner matrices.

SPE using C18 cartridge was also used for the extraction of pesticide residues (including atrazine) in waters samples from Brazil (Rocha et al. 2015). After extraction, pesticides were determined by LC-MS/MS. The results showed that C18 is not a good sorbent for atrazine; the LOQ was 5 µg/L and the recoveries obtained for water samples spiked at 20, 80 and 100 µg/L were between 109 and 127%.

Carabias et al. (2002) have compared C18 with two polimeric sorbents (Oasis and LiChrolut EN) for the extraction of atrazine, terbutryn and its metabolites, DEA, DEDIA, DEHA, DIHA, HA and HT. A HPLC-DAD was used for the separation, identification and quantification. The best results were obtained using LiChrolut EN polymeric cartridge (a hydrophobic polymer of styrene-divinylbenzene) when the elution was performed with methanol and ethyl acetate. The LODs obtained were between 0.1 µg/L for DIHA and DEDIA and 0.02 µg/L for the other analytes. The method was used to evaluate the presence of these herbicides and their degradation products in samples of surface and ground waters from agricultural zones of Salamanca and Zamora (Spain). Later, the authors developed a SPE method with LiChrolut EN sorbent coupled to a non-aqueous capillary electrophoresis separation with UV detection (NACE-UV) for the determination of ametryn, atrazine, prometryn, propazine, simazine and three main derivatives from the atrazine degradation products (DEA, DEHA and DIHA) (Carabias et al. 2006). The NACE-UV results were compared with those obtained with a HPLC-UV method. The results obtained show that both methods afford the same results in the analysis of surface and drinking water samples. The LODs in surface water samples were in the 0.04-0.32 µg/L and 0.11-1.2 µg/L ranges for the NACE-UV and HPLC-UV methods, respectively. The recoveries were significantly 100% in all cases.

Two cartridges, C18 and styrene-divinylbenzene (Bond Elut-ENV), and C18 disks were compared for the extraction of triazines (ametryn, atrazine, prometryn, propazine, simazine and terbutryn) and three degradation products (DEA, DEDIA and DIA) from water samples (Berzas-Nevado et al. 2007). The determination was carried out by GC-MS. The copolymer cartridge Bond Elut-ENV was selected because the recoveries obtained were higher than those obtained with reversed phases. The analytical performance characteristics were evaluated by means of a validation study on tap and deionised water samples. The recoveries were similar in deionised and tap water samples. The sorbent was adequate for DEA and DIA (recoveries nearly to 100%) obtaining worse recoveries for the others herbicides (between 67-76%, except for DEDIA with recoveries less than 15%).

Van Pinxteren et al. (2009) have tested two sorbents (C18 and Oasis HLB) for the determination of 10 pesticides (including atrazine, prometryn, propazine and simazine) in surface and ground water by HPLC-MS/MS. The comparison of both materials has showed that, in general, better recoveries were achieved with C18. The LODs obtained were in the low ng/L range (from 0.5 to 5.5 ng/L for the triazines studied). However, extraction yields were not very high (< 77%).

Different SPE cartridges (Oasis HLB 500 mg, Oasis HLB 150 mg, Oasis HLB 60 mg and Chromabond HR-X 60 mg) for the extraction of 33 pesticides including atrazine, DEA, DET, DIA, simazine and terbuthylazine in water using LC-MS/MS have been compared (Lissalde et al. 2011). The results have suggested that a simple hydrophobic polystyrene-divinylbenzene copolymer gave satisfactory and robust recoveries from mineral and river waters (between 85 and 105%) for all compounds.

Fingler et al. (2017) carried out a study of herbicide micropollutants (phenylureas, triazines and degradation products, chloroacetanilide, dinitroaniline) in surface, drinking and ground water in the area of Zagreb (Croatia). Triazines (atrazine and terbuthylazine) and degradation products

(DEA, DET and DIA) were extracted employing styrene-divinylbenzene sorbent cartridges and determined by HPLC-DAD. Atrazine and terbuthylazine were also extracted by SPE using C18 and analysed by GC-MS. The recoveries ranged from 81 to 106% with RSDs lower than 9%.

A SPE method employing Chromabond RP 200 mg prior to GC-MS analysis was applied for the determination micro-organic contaminants in groundwater in the city of Maribor (Slovenia) (Korosa et al. 2016). The pesticides and metabolites analysed were atrazine, DEA, DET, DIA, metolachlor, propazine, simazine and terbuthylazine. The most frequently detected compounds in groundwater were atrazine, DEA, DET and simazine.

3.3. Selective Sorbents

Complex samples contain large amounts of interfering species; therefore, there is a considerable interest in novel SPE materials with high selectivity. Thus, in recent years the use of molecularly imprinted polymers (MIPs), which have selective binding points capable of recognizing a specific molecule, has become one of the most important research fields in SPE. This type of sorbents has been employed for the extraction of triazines and their degradation products.

MIPs for atrazine, ametryn and irgarol were prepared by a multi-step swelling and polymerization method using ethylene glycol dimethacrylate as a cross-linker and methacrylic acid (MAA), 2-(trifluoromethyl) acrylic acid (TFMAA) or 4-vinylpyridine either as a functional monomer (Sambe et al. 2007). The MIP for atrazine prepared using MAA showed good molecular recognition abilities for chlorotriazine herbicides, while the MIPs for ametryn and irgarol prepared using TFMAA showed excellent molecular recognition abilities for methylthiotriazine herbicides. A restricted access media-molecularly imprinted polymer (RAM-MIP) for irgarol was prepared followed by in situ hydrophilic surface modification using glycerol dimethacrylate and glycerol monomethacrylate as hydrophilic monomers. The RAM-MIP was applied to selective pretreatment and enrichment of methylthiotriazine herbicides (ametryn, prometryn and simetryn) in river water, followed by HPLC-UV. The LOQs of ametryn, prometryn and simetryn were 50 ng/L. The recoveries of ametryn, prometryn and simetryn, at the LOQ level, were 95.6, 95.1 and 101% respectively.

A MIP for atrazine was synthesized by non-covalent method (Kueseng et al. 2009). The MIP provided high recovery and selectivity for atrazine while low recovery and selectivity were obtained for prometon, simazine and terbutryn. HPLC-DAD was used for triazines analysis. The recovery obtained using the MIP was higher than those achieved using two non-imprinted polymers, a commercial C18 and a granular activated carbon sorbent. The method provided high recoveries ranged from 94 to 99% at two spiked levels (0.8 and 8 µg/L) and the LOQ was 0.27 µg/L.

García-Galan et al. (2010) developed a selective and sensitive method for the simultaneous determination of triazines (atrazine, cyanazine, prometron, propazine, sebuthylazine and simazine) and degradation products (DEA, DET and DIA) in waters and soils. The method is based on sample extraction and pre-concentration with a commercial molecularly imprinted polymer cartridge (MIP4SPE Triazine 10) followed by LC-MS/MS analysis. The LOQs achieved were between 0.049-0.412 ng/L and 0.083-1.162 ng/L for the triazines in ground water and tap water respectively, except for cyanazine, which presented worse LOQs (7.9 and 4.97 ng/L for ground water and tap water respectively). The suitability of the method for the environmental trace analysis of triazine pesticides and their metabolites in aqueous samples was demonstrated through the analysis of several ground waters. The recoveries obtained using the MIP cartridges for the water matrices studied at two spiked levels were 56-128% for 2.5 ng/L and 31-105% for 10 ng/L.

The combination of a method based on the use of molecularly imprinted-solid phase extraction (MISPE) along with dispersive liquid-liquid microextraction (DLLME) has been reported for the selective extraction and pre-concentration of triazine pesticides from aqueous samples (Sorouraddin and Mogaddam 2016). Molecularly imprinted microspheres (template, atrazine) were used as sorbent in the SPE procedure for the extraction of atrazine and its analogs (ametryn, cyanazine, prometryn, propazine, simazine and terbutryn). The method employed 125 mg of microspheres and 12 mL of sample. Then the adsorbed analytes were eluted with 0.75 mL of methanol and mixed with 40 μL of carbon tetrachloride (as extraction solvent). In this process, the analytes were extracted into fine droplets of carbon tetrachloride and the fine droplets were sedimented in the bottom of the conical test tube by centrifugation. Finally, GC-FID and GC-MS were used for the separation and determination. The LODs and LOQs were between 0.2-7 and 0.5-20 μg/L, respectively. The relative recoveries obtained for atrazine in spiked samples were within in the range of 80-98%.

Another selective sorbent are mixed-mode ion exchange sorbents. Depending on the ion-exchange interactions established between the polymer and the analytes, there are two mixed-mode sorbents: strong cation-exchange (SCX) and strong anion-exchange (SAX). Although early mixed-mode sorbents were silica-based ion-exchange groups, nowadays polymer-based materials are most usual. The mixed-mode ion exchange sorbents used for the extraction of triazines and their major hydroxyl and dealkylated degradation products are Oasis MCX (SCX) and Oasis MAX (SAX). These mixed-mode sorbents are based on the Oasis HLB skeleton and further chemically modified with sulphonic groups for SCX and dimethylbutylamine for SAX.

Thus, Oasis MCX cartridges have been used for the simultaneous extraction of terbuthylazine and its five major degradation products (DET, desethyl-hydroxy-terbuthylazine (DEHT), DIA, DIHA and HT) (Papadopoulos et al. 2007). The determination was carried out by HPLC-DAD. The LOQ was found to be 0.2 μg/L for DIHA and 0.04 μg/L for the other compounds. Oasis MCX had shown good results for the degradation products studied (recoveries ranged from 70 to 80%).

A mixed-mode anion exchange sorbent was used to analyze 31 herbicides (including atrazine, cyanazine, DEA, DIA, HA, metribuzin, prometryn, propazine, simazine, terbuthylazine and terbutryn) in storm waters (Zhang et al. 2014). The SPE procedure was carried out using Oasis MAX cartridges, which are able to retain a wide range of herbicides with acidic-neutral-basic characteristics. The neutral and basic herbicides can be effectively eluted with methanol, after which the acidic herbicides can be eluted using acidified methanol. The determination was carried out by LC-MS/MS. The method achieved LODs of 1-30 ng/L for the triazines under study, with adequate recoveries (78-106%) except for DEA, DIA, HA, prometryn and terbuthylazine (65-79%).

Akdogan et al. (2013) developed a method for determination of atrazine, DEA, DIA and simazine in aqueous samples using Amberlite XAD-4 resin as sorbent. After extraction, the compounds were determined by HPLC-DAD. Good results were achieved; the LODs were between 0.084 and 0.121 µg/L and the recoveries ranged from 99.6 to 104.8%.

3.4. Other Sorptive Materials. Nanostructured Materials

Other sorbents such as carbon-based materials have also been used for the extraction of triazines in water samples. The main property of coal is its high specific surface area that entails a high adsorption capacity and strong interaction with the aromatic rings of organic molecules.

A method based on SPE using bamboo charcoal as sorbent and HPLC-UV for the determination of atrazine and simazine in waters was developed (Zhao et al. 2008). The results showed good sensitivity (LODs were 0.1 µg/L for both compounds). The method was applied to the analysis of tap water and well water samples and satisfactory recoveries were obtained (75-107% and 80-106% for simazine and atrazine respectively).

Graphene was used for dispersive solid-phase extraction of 11 triazine herbicides (ametryn, atrazine, cyanazine, cyprazine, metribuzin, procyazine, prometon, prometryn, propazine and terbuthylazine) and 5 neonicotine insecticides from tap and river water (Wu et al. 2015). Triazines were determined by GC-MS. The results indicated that graphene was an excellent sorbent for the analytes studied. The recoveries of the pesticides were 83.0-108.9% and the LODs ranged from 0.03 µg/L to 0.40 µg/L.

Others sorbents based on nanoparticles, nanotubes, nanofibres or nanowires for extraction and preconcentration of triazines in water samples are presented below. These materials show a great potential as sorbent for sample preparation with large specific surface area, providing more analyte interaction sites compared with traditional SPE sorbents.

Therefore, multiwalled carbon nanotubes (MWCNTs) has exhibited exceptional merit as SPE sorbents for enrichment of environmental pollutants. Zhou et al. (2006) described a sensitive method for the determination of atrazine and simazine using MWCNTs as solid phase sorbents followed by HPLC-DAD. The LODs of atrazine and simazine were 33 and 9 ng/L respectively. Good analytical performance was achieved from real water samples such as river water, reservoir water, tap water and wastewater. The spiked recoveries for both analytes were over the range of 82.6-103.7% in most cases.

MWCNTs disks were also used for preconcentration of atrazine and simazine in water samples followed by GC-MS analysis (Katsumata et al. 2010). The LODs obtained were 2.5 and 5.0 ng/L for atrazine and simazine, respectively. The proposed method was applied to the determination of atrazine and simazine in environmental water samples and the recoveries were in the range of 87 to 110%.

A micro-SPE technique was developed by fabricating a rather small package including a polypropylene membrane shield containing the appropriate sorbent for the extraction of atrazine, ametryn and terbutryn from aqueous samples (Bagheri et al. 2010). Various sorbents including synthesized aniline-ortho-phenylene diamine copolymer (PANI-OPD), synthesized polypyrrole (PPy), MWCNTs, C18 and charcoal were examined as extracting media. Conductive polymers exhibited better performance, although the efficiency of PANI-OPD was slightly lower than PPy being the latter selected as the most suitable sorbent for extraction of triazine herbicides from water samples. The micro-SPE-GC-MS offers enough sensitivity (LODs were in the range of 0.01-0.04 μg/L). The same authors developed a method based on microextraction in packed syringe (MEPS) combined with GC-MS for the multiresidue determination of triazines (ametryn, atrazine, terbutryn), organochlorine and organophosphorous pesticides in aqueous samples (Bagueri et al. 2012). A polyaniline (PANI) nanowires network has been synthesized and used as sorbent. Good results were obtained; the LODs of the triazines were in the range of 0.2-0.3 µg/L. The developed method was applied to the Zayandeh-rood river water samples and the recoveries obtained for the spiked real water samples for ametryn, atrazine and terbutryn were 73, 66 and 89% respectively.

Yang et al. (2015) have developed a method using a new SPE sorbent of Nylon 6/polypyrrole nanofibres mat (NFsM) and a new packing format of SPE disk for the determination of atrazine in tap and lake water samples by HPLC-UV. The LOD was 0.03 µg/L and the recoveries (ranged from 94.7 to 114.9%) were adequate.

A novel solid phase extraction sorbent based on carboxyl-modified polyacrylonitrile NFsM (COOH-PAN-NFsM) was fabricated and evaluated for the extraction of atrazine and its metabolites DEA and DIA from environmental water samples (Cao et al. 2017). The target analytes in 10 mL of water sample were extracted by only 4 mg of COOH-PAN-NFsM, eluted with 400 µL of methanol and analyzed by HPLC-DAD. The recoveries ranged from 81.4 to 120.3% and the LOQs were in the range 0.09-0.12 µg/L.

In recent years, a new SPE method using magnetic materials, known as magnetic dispersive solid phase extraction (MDSPE), has been increasing attention. MDSPE is performed under an external magnetic field without tedious centrifugation of filtration procedures. Magnetic sorbents dispersed into the solution can increase the contact interface with analytes significantly. Moreover, magnetic sorbents can be easily retrieved, which makes the sample pretreatment procedure more convenient, timesaving and economical. Pure inorganic magnetic particles, such as Fe_3O_4, could be easily aggregated and are not suitable to be used for extracting organic compounds; therefore, surface modification is necessary. Numerous types of natural or synthetic polymers, novel molecules and inorganic materials have been coated on the surface of the magnetic particles.

A SPE procedure using superparamagnetic Fe_3O_4 nanoparticles (NPs) as extracting agent was developed for the analysis of atrazine, prometryn, propazine, and terbutryn by HPLC-MS/MS in surface water (Song et al. 2007). The NPs showed an excellent capability to retain the compounds tested, and a quantitative extraction was achieved within 10 min under the testing conditions. After extraction, the superparamagnetic NPs were easily collected by using an external magnet. The elution of analytes retained on the Fe_3O_4 NPs was found very difficult, which was solved by dissolving Fe_3O_4 nanoparticles with a HCl solution. The extraction of small organic molecules such as triazine pesticides was effective. Theoretically, 100% analytes retained on NPs can be retrieved into solution by means of the proposed acid dissolution.

Triazines (atrazine, prometon, prometryn and propazine) were extracted from environmental water samples using graphene-based Fe_3O_4 magnetic NPs (G-Fe_3O_4 MNPs) followed by HPLC-DAD (Zhao et al. 2011). After the extraction, the sorbent can be conveniently separated from the aqueous samples by an external magnet. The LODs ranged between 0.025 and 0.040 µg/L. The method was applied to the analysis of the triazine herbicides in different water samples (lake, river and reservoir) and the recoveries were satisfactory (between 89.0% and 96.2%). Zhang et al. (2013) have also employed G-Fe_3O_4 MNPs as sorbent in magnetic SPE coupled with GC-MS for the determination of atrazine, cyanazine, metrybuzin, prometryn, propazine, simazine and simetryn in water samples. Low LODs (between 1 and 5 ng/L) and good recoveries (80-118%) in the water samples were achieved.

A novel sorbent based on polythiophene/chitosan magnetic nanocomposite was proposed for the preconcentration of triazines (ametryn, atrazine, and terbutryn) in aqueous samples prior to GC-MS (Feizbakhsh and Ehteshami, 2016). Polythiophene is an interesting conducting polymer and chitosan can be used as a good support for the growth of Fe_3O_4 nanoparticles, but also as a spacer for the inhibition of nanoparticles aggregation. The recoveries were between 96 and 102% and the LODs oscillated from 10 to 30 ng/L. Later, the authors modified the sorbent by using electrospinning polymer nanofibres of polyamide, polycarbonate and polyurethane. The results confirmed that magnetic polyamide nanocomposite showed higher extraction efficiency (Feizbakhsh and Ehteshami, 2017).

Metal-organic frameworks (MOFs) are a new class of hybrid inorganic-organic microporous crystalline materials self-assembled straightforwardly from metal ions with organic linkers via coordination bonds. Zhou et al. (2017) developed a magnetic nanoparticle supported zeolitic imidazolate framework-8 for the enrichment of triazine herbicides (ametryn, atrazine, prometryn and simazine) from fruits, vegetables and water. Instrumental analysis was conducted by UFLC and DAD detection. The recoveries were 88.0-101.9% with RSDs $\leq$ 8.8%.

Zhou et al. (2018) presented a surface imprinting strategy for the preparation of magnetic superhydrophilic molecularly imprinted composite resin based on MWCNTs (MWNTs@Fe$_3$O$_4$@MIR) to detect triazines in environmental water. The surface imprinting was carried out by one-pot condensation of resorcinol, formaldehyde and melamine on the surface of MWNTs@Fe$_3$O$_4$. The condensation introduces abundant hydrophilic functional groups (hydroxyl, amino and ether linkages) into the resin while the template was imprinted. The analysis was carried out by HPLC-MS/MS. The effect of the amount of template (ametryn) on the imprinting performance was investigated. The method was successfully applied to detect ametryn, atrazine, desmetryn, prometryn, propazine and simazine. The LOQs were in the range 0.022-0.227 µg/L and recoveries oscillated from 87% to 100%.

Nasrollahpour and Moradi (2018) developed a vortex assisted magnetic dispersive solid phase microextraction method using Fe$_3$O$_4$@MIL-100 (Fe) for the extraction of four triazines (ametryn, atrazine, prometon and simazine) in environmental water (lake, river and seawater). Under optimized conditions, 2 mg of sorbent was dispersed in 1 mL of water by sonication. The resultant suspension was injected into 50 mL of sample. The mixture was placed in a vortex mixer for 3 min to enhance the adsorption of triazines onto the Fe$_3$O$_4$@MIL-100 (Fe). Then the sorbent was separated from the solution under a strong magnetic field, 200 µL of dichloromethane were added and the mixture was sonicated for 3 min to desorb the retained triazines. The analysis was performed by

HPLC-DAD. The LODs and LOQs were in the range 6.1-15.7 µg/L and recoveries oscillated from 97.5% to 101.5%.

Ionic liquid-magnetic graphene (IL-MG) composite was used as sorbent in MDSPE to rapidly extract triazines (ametryn, atrazine and cyanazine) from surface water (Zhang et al. 2018). The IL and magnetic Fe_3O_4 nanoparticles act as spacers inserting between the layers of graphene to prevent the aggregation of graphene, which improves the adsorption ability because of the large specific surface area of IL-MG. The analysis was carried out by HPLC –UV Vis. The LOQs were in the range of 0.31-0.51 µg/L and the recoveries obtained were between 97.0 and 100.8%.

3.5. On-Line SPE

On-line SPE is an effective technique for the analysis of trace contaminants, such as drugs of abuse, pesticides, pharmaceuticals, and hormones, in a variety of matrices (water, urine and plasma). On-line SPE devices provide advantages over off-line SPE methods such as the minimal amount of solvents required for extraction, fast sample preparation and small sample volumes (Trenholm et al. 2009). However, on-line SPE has also disadvantages including the complexity of the valve-switching set-ups and operation, lack of flexibility as compared to off-line SPE, and possible matrix interferences from loading the entire extracted sample (Liska 1993). Most of these problems have been solved with advances in automated on-line SPE systems, integrated and flexible software programs as well as the use of MS/MS detectors.

During the last decade, the use of on-line SPE coupled to LC-MS has increased and several methods have been published for the analysis of triazines and their degradation products in water. The most employed sorbent for on-line SPE are cartridges filled with rigid macroporous spherical particles of polystyrene and divinylbenzene (PLRP-s). Thus, Brix et al. (2009) have used polymeric cartridges PLRP-s to develop an on-line SPE-LC-UV method for the determination of eight triazines (ametryn, atrazine, cyanazine, metrybuzine, prometryn, propazine, simazine and terbutryn) in drinking water. During the development of the method, it was observed that the retention times of ametryn, prometryn, and terbutryn in Milli-Q water were different from those in chlorinated Milli-Q water, indicating the formation of new products. The investigation of the products by UPLC-Q-TOF-MS/MS has shown that the three triazines follow a similar transformation pathway, forming four new molecules whose structure have been elucidated.

Postigo et al. (2010) have developed a methodology based on on-line SPE-LC-MS/MS for the determination of 22 media to highly polar pesticides in groundwater. Sample preconcentration was performed by using PLRPs cartridges for the analysis of 16 pesticides measured in the positive ionization mode (including atrazine, cyanazine, DEA, DIA, simazine and terbuthylazine). The methodology developed allows the determination of the triazines at low ng/L level (LOQs < 1.29 ng/L for all triazines) with satisfactory accuracy (recovery percentages higher than 77%). Köck et al. (2010 and 2012) applied this on-line SPE-LC-MS/MS method for the analysis of pesticides including atrazine, cyanazine, DEA, DIA, simazine and terbuthylazine in river water samples.

An on-line SPE-UPLC-MS/MS using PLRPs cartridges was also employed for the pre-concentration and analysis of 37 pesticides including atrazine, prometryn, propazine, sebuthylazine, simazine, terbumeton, and three degradation products (DEA, DET and DIA) in surface water samples (Hurtado-Sánchez et al. 2013). The validation parameters were performed and the LOQs for triazines were lower than 0.018 µg/L using an injection sample volume of 1.5 mL. The recoveries were evaluated at four concentration levels (0.01, 0.03, 0.10 and 0.20 µg/L) and acceptable values were found except for DEA at the lowest level (59%).

Recently, Cotton et al. (2016) developed and validated a multiresidue method for the analysis of more than 500 compounds (pesticides and drug residues) in water by the combination of on-line SPE-UPLC-HRMS. A SPE cartridge Oasis HLB was employed and a concentration factor of 500 up to 1000 was gotten. The method enables the simultaneous semi-quantitative analysis of 539 compounds in 36 min with only 5 mL of water. Six triazines (ametryn, atrazine, prometon, propazine, simazine and terbuthylazine) and four degradation products (DEA, DET, DIA and HA) were determined. The method was applied to the analysis of tap water from the Paris area. The pesticides more frequently found were atrazine and its metabolites, propazine and simazine. Although some pesticides were present in all samples, their levels were below the European regulatory level for tap water (0.1 µg/L).

4. MICRO-EXTRACTION TECHNIQUES

SPE is being replaced by fast micro-extraction techniques that minimize the waste of organic solvents according with the principles of Green Chemistry (Anastas et al. 2010). So, solid phase micro-extraction (SPME), stir bar sorptive extraction (SBSE), liquid-phase micro-extraction (LPME) and solid phase membrane tip extraction (SPMTE) have been applied for the extraction and preconcentration of triazines in water samples as an alternative to the SPE techniques. Regarding SPME, Mohammadi et al. (2009) employed a Headspace SPME method with dodecylsulfate-doped polypyrrole fibre coupled to ion mobility spectrometry for the determination of atrazine and ametryn in soil and water samples. The LODs for ametryn and atrazine in water were 10 and 15 µg/L respectively. Passeport et al. (2010) developed a SPME-GC-MS multiresidue method for 16 pesticides, including atrazine, with good LOQ (0.05 µg/L for atrazine) but inadequate recovery (143 ± 13.6%).

A SPME fibre was produced by copolymerization of methacrylic acid-ethylene glycol dimethacrylate imprinted with atrazine for the selective extraction and analysis of triazine herbicides (Djozan and Ebrahimi 2008). At the optimum conditions the fibre is firm, inexpensive and thermally stable up to 280 ºC, which is of vital importance in SPME coupled with GC or GC/MS. High extraction efficiency was obtained for ametryn, atrazine, cyanazine, prometryn, propazine, simazine and terbutryn with LODs of 69, 20, 81, 68, 80, 70 and 88 µg/L respectively. The reliability of prepared fibre for the extraction of the triazines in real samples was investigated in spiked tap water samples with satisfactory recoveries (between 96.3 to 99.6%). The same authors have also fabricated another SPME fibre through ultra violet irradiation polymerization of ametryn-molecularly imprinted polymer on the surface of anodized-silylated aluminium wire (Djozan et al. 2010). This fibre showed high selectivity with great extraction capacity towards triazines. SPME coupled with GC with flame ionization detector (FID) was used for the analysis of ametryn, atrazine, cyanazine, prometryn, propazine simazine and terbutryn. The LODs (between 9 to 85 µg/L) were very high and the recoveries from tap water samples spiked at 100 and 500 µg/L were satisfactory, although, it should be into account the high levels employed to spike the samples. In a subsequent study, the same authors determined triazines by SPME using a inside-needle extraction method through MIP (atrazine as template) on the internal surface of a stainless steel hollow needle, which was oxidized and silylated (Djozan et al. 2012). GC-FID was used for the analysis. The extraction of ametryn, atrazine, cyanazine, prometryn, simazine and terbutryn resulted in LODs between 2.6 and 42 µg/L. The recoveries of triazine compounds from tap and groundwater samples spiked at 500 and 1000 µg/L were satisfactory, although, it should be into account the high levels used to spike the samples.

Furthermore, a SPME using graphene-coated fibre coupled to HPLC-DAD was applied for the determination of ametryn, atrazine, prometon and prometryn in water samples (Wu et al. 2012). The LODs were between 0.05-0.2 µg/L and the recoveries (at spiking levels of 20 and 50 ng/mL) ranged from 86.0 to 94.6%. The graphene-coated fibre compared with two commercial fibres (CW/TPR and PDMS/DVB) showed higher extraction efficiency.

Zhang et al. (2017) fabricated a new multiple monolithic SPME fibre using a polydopamine-based monolithic for the determination of triazine herbicides (atraton, atrazine, prometryn, simazine, terbumeton and terbuthylazine) in environmental water samples. After liquid desorption, the analysis was carried out by HPLC-DAD. The results indicated that the LOQs for the target compounds were between 0.10-0.45 µg/L. The developed method was applied to the analysis of the triazine herbicides in different water samples (lake, river, and farmland waters). The recoveries ranged from 79.6 to 117%.

Moreover, in tube-solid phase microextraction (IT-SPME) coupled to miniaturized liquid chromatography is attractive mainly due to the column efficiency improvement, sensitivity enhancement and reduction of solvent consumption. This procedure integrates the online extraction and preconcentration. Good sensitivities are achieved although the main limitations is the low extraction efficiency. The most widely used capillaries are segments of open capillary columns, generally commercial GC columns. Recently, De Toffoli et al. (2018) have developed an on-line IT-SPME coupled to HPLC-MS/MS for the determination of ametryn, atrazine and simazine in water samples. The method was based on the employment of a packed column containing graphene oxide (GO) supported on aminopropyl silica (Si). The method showed satisfactory results with low detection limits (1.1-2.9 ng/L) and high extraction efficiency.

SBSE has also been employed for the extraction and preconcentration of triazines in water samples. Thus, polyurethane foams were applied as polymeric phases for the determination of seven triazines (ametryn, atrazine, prometon, prometryn, propazine, simazine and terbutryn) in ground and superficial waters by SBSE-HPLC-DAD. The methodology provided poor recoveries (20.4-62.0%) and LODs between 0.1 and 0.5 µg/L (Portugal et al. 2008).

Roldán-Pijuán et al. (2015) developed stir fabric phase sorptive extraction (SFPSE), which integrates sol-gel hybrid organic-inorganic coated fabric phase sorptive extraction media with a magnetic stirring mechanism. The microextraction device has been analytically evaluated using seven triazine herbicides (atrazine, prometryn, propazine, secbumeton, simazine, terbumeton, and terbutryn). Two flexible fabric substrates, cellulose and polyester, were used as the host matrix for three different sorbents (sol-gel poly (tetrahydrofuran), sol-gelpoly (dimethyldiphenylsiloxane and sol-gel poly (ethylene glycol). Triazines were better extracted by sol-gel poly (ethylene glycol). The LODs, using LC-MS/MS for the analysis of the triazine herbicides, were in the range of

15-26 ng/L. The developed method was applied for the determination of selected triazines from three river water samples. Poor recoveries of the target analytes, in the range from 75 to 126%, were found.

Liquid-phase microextraction is simple, rapid, effective and low cost extraction method that minimize the toxic organic solvents used to extract the analytes. It is usually classified into three main types: single-drop microextraction (SDME), hollow-fibre liquid-phase microextraction (HF-LPME) and dispersive liquid-liquid microextraction (DLLME). These extraction methods have been used for the extraction of triazines in waters.

Ye et al. (2007) developed a method to determine atrazine, cyanazine and simazine by SDME in tap and reservoir waters with LODs ranged from 0.03 to 0.06 µg/L. The proposed SDME method was applied for the analysis of atrazine, cyanazine and simazine in four environmental water samples (tap water, well water, reservoir water and snow). The method showed poor recoveries and high RSD (12-15%) were observed for all compounds in well water.

A HF-LPME method has been employed for the analysis of 16 pesticide (including atrazine, propazine and simazine) in natural water samples (Trtic-Petrovic et al. 2010). A HPLC-MS/MS was used for the determination of pesticides. The LOQs of the triazines under study ranged from 0.087 to 0.156 µg/L and good recoveries were obtained.

Recently, Yang et al. (2018) developed a HF-LPME method combining on-line sweeping micellar electrokinetic chromatography with UV detection for the determination of seven triazines (ametryn, atrazine, prometon, prometryn, propazine, simazine and terbutryn) in honey, tomato and environmental water samples. The LODs were in the range of 0.07-0.69 µg/L with recoveries from 85.2% to 114%.

Nagaraju et al. (2007) have determined eight triazines (atrazine, desmetryn, prometryn, propazine, sebuthylazine, secbumeton, simazine and simetryn) in river and tap water samples using DLLME and GC-MS. The LODs were between 0.021 and 0.12 µg/L for most of the analytes and the recoveries of triazines at a spiking level of 5.0 µg/L were between 85.2-119.4%. The method was compared with SPME and HF-LPME being

DLLME faster and showing high enrichment factors and recoveries for the extraction of triazines from water.

DLLME has also been used to determine atrazine and simazine in tap, ground and reservoir water using HPLC-UV detection (Zhou et al. 2009). The LODs were 0.1 and 0.04 µg/L for atrazine and simazine, respectively. The proposed method was applied to the analysis of real water samples and good results were achieved for reservoir water with recoveries of 90% and 91% for atrazine and simazine respectively. However, poor recoveries (60 to 85%) were obtained for tap and ground waters.

DLLME coupled with HPLC-DAD for the determination of five triazines (ametryn, atrazine, prometon, prometryn and simazine) in water and soil samples has been studied (Wang et al. 2011). The LODs were 0.05-0.1 µg/L for the water samples. In order to validate the accuracy of the proposed method reservoir, river and well water samples were spiked with the standards of the target analytes at the concentrations of 5 and 50 µg/L and the recoveries were satisfactory (between 84.2% and 102%).

Sanagi et al. (2012) developed a DLLME method based on solidification of floating organic droplet (DLLME-SFO) coupled with GC-MS for the analysis of atrazine, cyanazine, secbumeton and simazine in lake and tap water. The method gave low LODs (8-37 ng/L) and excellent relative recoveries (100-109%).

Recently, Bolzan et al. (2015) have also developed a method for the extraction and preconcentration of different classes of pesticides (including atrazine and simazine) in mineral water samples by DLLME. The LODs for atrazine and simazine were 0.05 µg/L and recoveries 120%, with RSDs between 9 and 10%.

Moreover, variations and modifications of LPME have been carried out as dispersive liquid-phase microextraction (DLPME), liquid-liquid-solid microextraction (LLSME) and hollow fibre-liquid-solid microextraction (HF-LSME). In this way, an ionic liquid based on dispersive liquid-phase microextraction (IL-DLPME) method for the determination of three triazine and two phenylurea herbicides in water samples has been developed (Wang et al. 2010). The extracts were analysed by HPLC-DAD. Poor recoveries were obtained for the triazines in water samples (67, 77 and 54% for atrazine, prometryn and simazine respectively). The LODs for the three triazines were between 0.46 and 0.89 µg/L.

Hu et al. (2009) have proposed a LLSME technique based on porous membrane-protected MIP-coated silica fibre. In this technique, a MIP coated silica fibre was protected with a length of porous polypropylene hollow fibre membrane, which was filled with water-immiscible organic phase. Subsequently the whole device was immersed into aqueous sample for extraction. The LLSME technique was a three-phase microextraction approach. The target analytes were firstly extracted from the aqueous sample through a few microliters of organic phase residing in the pores and lumen of the membrane, and finally extracted onto the MIP fibre. The LLSME method coupled with HPLC-UV was applied to the analysis of ametryn, atrazine, propazine, simetryn, terbuthylazine and terbutryn in complex aqueous samples (sludge water, watermelon, milk and urine) with low LODs (6-20 ng/L) and satisfactory recoveries.

Chen et al. (2014) have used a HF-LSME procedure for the microextraction of triazines (ametryn, atrazine, 2-amino-4-methoxy-6-methyl-1, 3, 5-triazine and terbuthylazine). Firstly, a MIP was prepared in a silica capillary mould by microwave irradiation. After that, the resulting monolithic bar was embedded in a porous hollow fibre membrane tube, in which a thin supported liquid membrane was formed in the pores and an acceptor solution was filled. The MIP-HFLSME-HPLC-UV method was successfully applied to detect the analytes in lake water. The results obtained were adequate (LODs between 0.18 and 0.35 µg/L and recoveries in the range of 72.8-113.2%).

Finally, SPMTE was also used for determination of triazines in water samples. See et al. (2010) have developed a method using MWCNTs as sorbent enclosed in SPMTE device and a micro-LC system for the determination of atrazine, cyanazine, propazine and simazine. Low LODs (1.2-1.8 µg/L) and satisfactory recoveries (95–101%) from river water samples were obtained.

A SPMTE using GC-MS method was also developed and validated for the analysis of triazine herbicides (atrazine and secbumeton) in stream and lake waters (Sanagi et al. 2015). The solid-phase membrane tip extraction was carried out in semiautomated dynamic mode on MWCNTs enclosed in a cone-shaped polypropylene membrane cartridge. The recoveries of atrazine and secbumeton were 88.0 and 99.0% respectively with LODs < 0.47 µg/L.

Furthermore, some of these techniques have been used for the extraction of triazines and some of their main degradation products such as IT-SPME, SBSE and dispersive microextraction solid phase extraction (DMSPE). Thus, Moliner-Martínez et al. (2015) describe a method for the determination of polar triazines (atrazine, propazine and terbuthylazine) and some degradation products (DET and HT), which combines on-line IT-SPME and LC-DAD. Different extractive coatings were evaluated: a capillary column with a polydimethylsiloxane (PDMS) coating, the same coating modified with carboxylated single-wall carbon nanotubes (c-SWCNTs) and carboxylated multiwall carbon nanotubes (c-MWCNTs). The results obtained showed that the immobilization of c-SWCNTs and c-MWCNTs on PDMS columns did not introduce considerable improvement for the extraction and preconcentration of triazines in the IT-SPME methodology. The LODs were in the 0.02-0.1 µg/L range.

Serra-Mora et al. (2017) also applied this technique to the determination of polar triazines and degradation products (atrazine, DET, DIA, HT, propazine and terbuthylazine) in waters and recovered struvite. Different extractive phases such as TRB-5, TRB-5/c-SWNTs, and TRB-5/c-MWNTs capillary columns were tested and a new polar-coated capillary, based on tetramethyl orthosilicate and trimethoxymethylsilane containing SO_2 nanoparticles TEOS-MTEOS-SiO$_2$NPs, was studied for the first time for the target compounds. The polar capillary column was proposed because it improved the retention of polar compounds and achieved better LODs (0.025-0.5 µg/L). They also compared the performance of IT-SPME coupled to nano LC (NanoLC) and IT-SPME coupled to capillary LC (CapLC) with similar configurations (one pump and a six-port injection valve) and detection by DAD. Sensitivity of NanoLC was 10-25 times higher using the same extractive phase.

Sánchez-Ortega et al. (2009) have developed a method based on SBSE coupled to GC-MS for the determination of atrazine, isometioazine, metamitron, metribuzine, simazine, simetryn, terbumeton, terbuthylazine and atrazine metabolites (DEA and DIA) in aqueous solution with excellent sensitivity (LOQs between 0.7 ng/L and 11.3 ng/L). The recoveries obtained in the analysis of spiked ground water samples ranged from 94.4 to 106.0%.

A dispersive micro solid-phase extraction (DMSPE) using single-walled carbon nanohorns (SWNHs) as sorbent was evaluated for the extraction of ten triazines (atrazine, desmetryn, prometon, prometryn, propazine, secbumeton, simazine, simetryn, terbumeton, and terbutryn) and one degradation product (DET) in water samples (Jiménez-Soto et al. 2012). The water samples containing the triazines were placed in a glass vial, 1 mL of the SWNHs dispersion was added and the mixture was stirred. Then, the SWNHs enriched with the analytes were recovered by filtration, the analytes were eluted with methanol and analysed by GC-MS. Satisfactory results were obtained in terms of sensitivity and recovery. The LODs ranged from 0.015 to 0.1 µg/L and the recovery in different water samples provided average values between 87% and 94%.

Finally, a DMSPE method based on a polymer cation exchange material was also applied for the determination of a total of 30 triazines (26 triazines and 4 degradation products) in drinking waters by UPLC-MS (Chen et al. 2015). The method achieved LODs of 0.2-30 ng/L with recoveries in the range of 70-112%.

These new techniques have some advantages over SPE, but they also have drawbacks such as low sensitivity for the triazines studied (Mohammadi et al. 2009, Djozan et al. 2008, 2010 and 2012), poor recoveries (Bolzan et al. 2015, Passeport et al. 2010, Portugal et al. 2008, Roldán-Pijuán et al. 2015, Wang et al. 2010, Ye et al. 2007, Zhou et al. 2009), extraction of few compounds (Sanagi et al. 2015, See et al. 2010) and, in many cases, they are very laborious (Hu et al. 2009, Trtic-Petrovic et al. 2010, Jiménez-Soto et al. 2012, Sanagi et al. 2012, Chen et al. 2014).

5. ABOUT OUR RESEARCH

The research described below was carried out as a part of several environmental projects. "Study of the implementation of the Marine Water Framework Directive in Galicia" funded by Galicia Government and "Program of Consolidation and Structuring of Units of Competitive Investigation of Galicia University System" potentially cofunded by European Regional Development Fund (ERDF) in the frame of the operative Program of Galicia 2007-2013. "Towards an advanced and environmentally responsible analytical methodology for the determination of emerging pollutants. Focused on the water cycle and water treatment processes" and "The impact of microplastics and emerging pollutants in marine ecosystems and the establishment of environmental quality criteria for these compounds" funded by Spanish Government. In these projects, development and application of methods for the determination of persistent and emerging pollutants in the different matrices of the marine environment (water, sediments and biota) have been carried out in collaboration with the Spanish Institute of Oceanography (IEO).

The marine environment is a valuable medium that has to be protected, conserved and rehabilitated in order to keep clean oceans and seas, being healthy and productive. After the application of triazine herbicides in soil, fluxes of these compounds can reach the marine environment through hydrological cycle. Once there, they are distributed in all compartments (water, sediments and biota) depending on the properties of the marine ecosystem and the nature of the sorbent phases. Consequently, further studies about the behaviour, effects and occurrence of triazines and their main degradation products in the marine environment are required. It should be note that Galicia is the highest producer of mussels in Europe and the fish and shellfish industry is very important in this area, exporting their products around the world. Therefore, the study of levels of these contaminants on fish and fishery products is of a great economic and environmental importance for Galicia. For this purpose, it is necessary to develop accurate, effective, simple and fast analytical methods that allow the determination of these compounds at low concentration levels.

In our research, analytical methodology for the simultaneous determination of nine triazines (ametryn, atrazine, cyanazine, prometryn, propazine, simazine, simetryn, terbuthylazine and terbutryn) and eight of their main degradation products (DEA, DEDIA, DEHA, DET, DIA, DIHA, HA and HT) in different compartments of the marine ecosystem (seawater, sediments and biota) has been developed. The proposed methods have improved the previous methodologies found in the literature, are sensitive, selective and simple, besides complying with Green Chemistry principles. Once analytical methods were validated, a sampling of seawater, sediments and aquatic biota (algae, fish, and molluscs) was held. It should be noted that from conventional to more sophisticated and novel analytical techniques have been employed for the determination of the target compounds. These techniques have been used for the first time either for the analysis of these compounds or for the matrices studied.

Concerning analytical techniques, three chromatographic methods were optimized and validated: HPLC-DAD, LC-MS/MS and UPLC-MS/MS.

At the beginning, a HPLC-DAD method was developed for the determination of nine triazines, employing C18 as stationary phase and gradient elution with acetonitrile-water. An adequate separation for all the compounds was achieved (Rodríguez-González et al. 2013). The absorbance was measured continuously in the 200-400 nm range and peaks areas quantification was carried out at 222.7 nm in order to achieve maximum sensitivity. It is important to note that there were no references in the literature, which achieved a good resolution of the nine triazines; in addition, in most of the articles, cyanazine and simazine remained total overlapped when acetonitrile was employed in the mobile phase.

Subsequently a LC-ESI-MS/MS method for the determination of seven triazines using MRM (multiple reaction monitoring) as acquisition mode, C18 as stationary phase and gradient elution employing acetonitrile-water was developed (Rodríguez-González et al. 2015a). Afterwards, a LC-ESI-MS/MS method for the determination of nine triazines and eight of their main degradation products was optimized (Rodríguez-González et al. 2017a). Since degradation products are strongly dependent on pH, the use of a mobile phase with a buffer or a modifier is necessary. For this reason, different modifiers were studied, and the best chromatographic separation was obtained using acetonitrile-acetic acid solution. For both methods, the optimization of MS/MS conditions included the optimization of cone voltage and collision energy for each pesticide in order to obtain the precursor ion and the corresponding products was carried out. It is noteworthy that no references have been found devoted to the simultaneous determination of these seventeen compounds.

Finally, as far as the instrumental techniques concerned, an UPLC-MS/MS method for the determination of nine triazines and eight of their degradation products using C18 bonded to ethylene-bridged hybrid (BEH) particles as stationary phase and gradient elution employing methanol-water with 5 mM ammonium acetate as modifier in both mobile phases was optimized. Respect to conventional liquid chromatography, the use of short packed columns with small particle size has advantages such as significantly shorten the analysis time as well as solvent consumption, higher efficiency of the chromatographic peaks and improvements in the LODs. On the other hand, it was the first time that this technique was used for the simultaneous determination of these seventeen compounds (Rodríguez-González et al. 2016).

Regarding extraction methodologies, several analytical methods have been optimized and validated for the determination of the target compounds in seawater, sediment and biota (seaweed, fish and shellfish). With respect to seawater, three analytical procedures have been developed: SPE off-line, SPE on-line and DLLME. The three methods achieved the Environmental Quality Standards (EQS) established by Legislation in force at the moment of the study.

In the first study, a SPE off-line method for the determination of nine triazine herbicides employing DAD as detection system was developed (Rodríguez-González et al. 2013). The objective of this work was to study the possible contamination of the Arousa and Vigo estuaries by triazines. The health of both estuaries is a priority for the government of the region because they are engaged in shellfish and fishing. The method consists on the extraction and preconcentration of 50 mL of water sample through

Oasis HLB cartridges, rinsing with 20 mL of Milli-Q water and elution with 3 mL of acetone. The validation and applicability of the method was studied for surface waters (river and seawater) showing good recoveries (> 93% for all compounds). The LOQs (between 0.46 and 0.98 µg/L) were adequate allowing the determination of these compounds at the levels required by the 2008/105/EC Directive (Legislation in force at the moment of the study). Furthermore, several aspects concerning sample pre-treatment, such as the sample filtration step and the stability of herbicides on the cartridge, were also evaluated. It could be concluded that the integrity of the analytes at -18°C during three weeks was not affected. Regarding filtration step, glass microfibre filters are recommended because an important decrease on recoveries of simetryn and terbutryn were observed when nylon membrane filters were employed. It is noteworthy that although SPE is a conventional technique, it is a good alternative since the use of cartridges allows the storage of the triazines until analysis, avoiding the problem associated with the maintaining herbicides integrity in aqueous solution when long periods are required before the analysis. It is noteworthy that methods based on solid-phase extraction combined with liquid chromatography had been commonly used to measure triazines in drinking and ground waters; however, there were not studies in seawater at the moment of the study.

Subsequently, the method developed for the determination of triazines in surface waters was adapted to analyse drinking water (Beceiro-González et al. 2014). The SPE-HPLC-DAD method employing 500 mL of water sample allowed the quantification of the target compounds at the levels required by European legislation for drinking waters. Oasis HLB sorbent demonstrated to be appropriate since it favourably extracted the nine triazine herbicides in tap and mineral water samples with an adequate accuracy for all compounds; free chlorine present in tap water did not affect the determination of the studied herbicides. The proposed method showed good sensitivity with LODs (between 0.010 and 0.023 µg/L) lower than 30% of parametric value requested by the Directive 98/83/EC, concerning the quality of water for human consumption. A second objective of this work was the employment of the method as a laboratory experiment for undergraduate students. For this purpose, a laboratory experiment for the students, which allows them to participate actively in learning of concepts through practical experience was designed.

Since the positive result of this experience, another work was done focused on teaching activity (González-Castro et al. 2016). The project consisted on the determination of four triazines in seawater samples by SPE-HPLC-UV using Oasis HLB cartridges. The main pedagogical goal of this experiment is the learning both the use of the instrument and the principles involved on the chromatographic analysis. Furthermore, it permits the undergraduate students of Chemistry Degree to gain experience on a number of essential techniques in the laboratory of Analytical Chemistry.

The following research carried out was the development of an environmental friendly, simple, fast and sensitive method for the determination of triazine herbicides in seawater based on DLLME employing DAD detection and confirmation of the results by ESI-MS/MS detection (Rodríguez-González et al. 2015a). Under optimized conditions, 25 mL of seawater sample were extracted with 300 μL of octanol as extractant solvent by shaken at 1200 rpm for 10 minutes. All the triazines exhibited good recoveries (between 81-102%) and the LOQs (0.19 and 1.12 μg/L), enabling the determination of these pollutants at the levels required by European Union legislation (Commission Directive 2008/105/EC). It is important to note that the method employs small volume of organic solvent and an agitation step instead of dispersive solvent, which simplifies the experimental procedure. However, it has as disadvantage that it cannot be applied to the determination of prometryn and terbutryn, neither to degradation products.

Afterward, the SPE method developed for the determination of triazines in surface waters was modified with the aim of analysing simultaneously triazines and their main degradation products employing, in addition, the detector MS/MS (Rodríguez-González et al. 2017a). For this purpose, sample volume was reduced (10 mL) due to the high sensitivity of mass detector, and milli-Q volume used for the washing step was also reduced to avoid losses of the more polar metabolites. Furthermore, since a concentration step was not necessary, elution was carried out with 2.5 mL of methanol. Under the optimum conditions, the proposed methods provided adequate limits of quantification (between 0.05-0.45 µg/L) and adequate recoveries for all compounds (87.5-99.4%). It is noteworthy that methods based on solid-phase extraction combined with LC-MS/MS had been used to measure triazines and some degradation products in river waters; however, there were not studies in seawater at the moment of the study.

Finally, an on-line SPE method coupled with UPLC-MS/MS was developed for the quantification and confirmation of 17 compounds (nine triazine herbicides as well as their main degradation products) in seawater (Rodríguez-González et al. 2016). The extraction of the compounds from seawater was performed by loading 5 mL of seawater through an Oasis HLB on-line SPE cartridge. After sample loading, the compounds were eluted to the LC column with the chromatographic mobile phase. The method has shown suitable precision and good recovery values for all compounds. The LOQs enabled the determination of these pollutants at the levels required by European Union legislation (Commission Directive 2013/39/EU) using only 5 mL of sample. Consequently, it can be an important tool to control the presence of triazines and their degradation products at trace levels in seawater samples in compliance with EU directives. On-line SPE-UPLC-MS/MS method showed to be a fast, sensitive and robust alternative method to traditional off-line SPE for the analysis of nine triazines and their main degradation products. The proposed method is suitable to be used in routine analysis due to sample pre-treatment is not required and allows rapid trace enrichment from low sample volume with minimal sample handling. Furthermore, on-line SPE reduces chemical waste due to the use of a minimal amount of extraction solvents. It is noteworthy that a few methods based on solid-phase extraction combined with UPLC-MS/MS have been used to measure triazines and degradation products in river waters; however, there were not studies in seawater. Furthermore, an important difference of the proposed method with previously described methodology for the analysis of triazines herbicides and their main degradation products is the determination of a major number of degradation products simultaneously with triazines.

As regards to solid matrices, analytical methodology based on matrix solid phase dispersion (MSPD) using DAD as detection system has been optimized for the determination of the nine triazines in sediments, seaweeds, fish and shellfish. Furthermore, in the case of sediments, degradation products were also analysed employing MS/MS as detection system. Because of the different nature and complexity of the several matrices, it was necessary to optimize for each one the sample amount and the type and amount of dispersing agents, co-sorbents and elution solvents.

MSPD technique has been widely used for the extraction of different pesticide residues from biological tissues, with vegetal and/or animal origin; however, it is noteworthy that references for the determination of triazines by MSPD are still scarce and furthermore, in the most cases, few triazines are included in these studies. Regarding environmental matrices, such as soils or sediments, studies of triazines employing this technique are limited and recent, and to the best of our knowledge, there were only two references devoted to the determination of triazines employing this extraction technique at the moment of the study.

Once the analytical methods were optimised, all of them were validated in terms of linearity, sensitivity, precision and accuracy according to validation parameters and criteria from SANCO Guidelines. The obtained results were highly satisfactory for all developed methodologies, indicating that all methods meet the requirements stipulated in compliance with EU legislation.

The determination of triazines in seaweeds was the first study carried out. Under optimized conditions, the procedure consists on blending 1 g of green seaweed in a mortar employing octasilyl-derivatized silica (C8) as dispersing agent (2 g) and subsequent transferring into a SPE column cartridge containing Envi-CarbII/PSA (0.5 g/0.5 g) as clean-up co-sorbent. Then the dispersed sample was washed with 10 mL of hexane and triazines were eluted with 20 mL of ethyl acetate and 5 mL of acetonitrile. The reliability of the method was evaluated in terms of recovery by spiking a red seaweed and a brown one. The obtained results demonstrated that the method achieved satisfactory recoveries for all compounds, which indicates that this procedure could be established as a suitable methodology for routine analysis to screen and monitor triazines in different types of seaweeds in compliance with EU legislation (Rodríguez-González et al. 2014). It is worthy to note that studies of triazines in seaweeds are very scarce and recent, and no studies using MSPD had been done to extract triazines from seaweeds.

Regarding shellfish, two dispersing agents (octadecylsilyl bonded silica (C18) and florisil) and eight clean up co-sorbents (florisil, silica, silica/alumina, Envi™ Carb, Envi-Carb-II/PSA, SAX/PSA, Envi-Carb-II /SAX/PSA and C18) were assayed. The best results were obtained with C18 as dispersing agent and Envi-Carb-II/SAX/PSA as clean-up co-column. Under final working conditions, 0.5 g of mussel was homogenized with 2 g of C18 and then the final mixture was transferred into a SPE column containing a triple sorbent layer of Envi-CarbII/SAX/PSA (0.5 g/0.5 g/0.5 g) (Rodríguez-González et al. 2015b). Once packed MSPD columns, the washing and elution steps were performed following the procedure previously described for seaweed samples. The method provided satisfactory accuracy and precision for the determination of triazines in mussels. Subsequently, this method was adapted and validated for the determination of triazines in trout samples. Because of the higher lipid content in aquaculture trout than in mussel, it was necessary to reduce the sample amount to 0.2 g (González-Castro et al. 2015). It is important to highlight that the suitability of a procedure based on MSPD for the extraction of these chemicals residues from fish and shellfish was demonstrated for the first time.

Concerning to sediments, three dispersing agents (C18, graphitized carbon black (GCB) and diatomaceous earth) and four clean-up co-sorbents (Florisil, GCB, primary and secondary amine (PSA)/SAX, and GCB/PSA) were assayed. The best results were obtained with GCB as dispersing agent without co-sorbent, using 20 mL of ethyl acetate as elution solvent. Under optimized conditions, 1 g of sediment sample was blended with 1 g of ENVI-Carb in a glass mortar with a pestle for 5 min. The final mixture was transferred into a glass syringe and once packed, elution was performed with 20 mL of ethyl acetate (Rodríguez-González et al. 2017b). The developed method provides satisfactory trueness and precision for the determination of triazines in marine sediments. Subsequently, the MSPD method was modified with the aim of analysing simultaneously triazines and their main degradation products employing, in addition, the detector MS/MS (Rodríguez-González et al. 2017a). For this purpose, due to the higher polarity of the degradation products, the eluent was slightly modified adding 5 mL of acetonitrile to the eluent. To the best of our knowledge there was only one reference in the literature devoted to the determination of triazines by MSPD in sediments; furthermore, this study only included one triazine (atrazine) and none degradation product.

Finally, the developed methods were applied to the analysis of samples. The SPE-HPLC-DAD method was applied to analyse fifty seawater samples collected at two different estuaries (Arousa Island estuary and Vigo estuary) from Galicia (NW Spain). Both areas are widely dedicated to shell fishing and fishing. The herbicides under study have not been detected in the samples analysed.

The DLLME-HPLC-DAD, MSPD-HPLC-DAD, SPE-LC-ESI-MS/MS and MSPD-LC-ESI-MS/MS were applied to the analysis of the target compounds in seawater and sediments samples collected from ten points susceptible to contamination by triazines from estuary of a Coruña (Galicia, NW Spain). Although only one sample of sediment contained detectable amount of terbuthylazine, the analysis of these compounds is of high importance in order to generate information related to their levels in areas where seafood is growing (mussels, crabs, oysters…) because of their great economic and environmental interest. Furthermore, the Galicia seafood industry is very important, exporting its products around the world; therefore, this study is of a special relevance.

The on-line SPE-UPLC-MS/MS method was applied to the analysis of triazines and their degradation products in ten seawater samples collected from ten beaches susceptible to contamination by triazines in the seashore of Matosinhos (North of Portugal). Although the triazines and degradation products under study have not been detected in the samples analysed, the monitoring of their levels in marine ecosystems situated close to areas of intensive horticulture is of great local interest both economic and environmental.

Regarding biota samples, the MSPD-HPLC-DAD methods were applied to different edible samples of seaweeds and mussels purchased from local markets of A Coruña. Although the triazines under study were not detected in the samples, the analysis of these compounds in biota is of great interest in order to evaluate risks for human health and also to control the quality of the marine environment.

REFERENCES

Akdogan, A; Divrikli, U; Elci, L. Determination of triazine herbicides and metabolites by solid phase extraction with HPLC analysis. *Analytical Letters,* 2013 46, 2464-2477.

Anastas, P; Eghbali, N. Green Chemistry: Principles and Practice. *Chemical Society Reviews,* 2010 39, 301-312.

Andreu, V; Picó, Y. Determination of pesticides and their degradation products in soil: Critical review and comparison of methods. *Trends in Analytical Chemistry,* 2004 23, 10-11.

Bagheri, H; Khalilian, F; Naderi, M; Babanezhad, E. Membrane protected conductive polymer as micro-SPE device for the determination of triazine herbicides in aquatic media. *Journal of Separation Science,* 2010 33, 1132-1138.

Bagheri, H; Alipour, N; Ayazi, Z. Multiresidue determination of pesticides from aquatic media using polyaniline nanowires network as highly efficient sorbent for microextraction in packed syringe. *Analytica Chimica Acta,* 2012 740, 43-49.

Barbosa, MO; Ribeiro, AR; Pereira, MFR; Silva, AMT. Eco-friendly LC-MS/MS method for analysis of multi-class micropollutants in tap, fountain, and well water from northern Portugal. *Analytical and Bioanalytical Chemistry,* 2016 408, 8355-8367.

Beceiro-González, E; González-Castro, MJ; Pouso-Blanco, R; Muniategui-Lorenzo, S; López-Mahía, P; Prada-Rodríguez, D. A simple method for simultaneous determination of nine triazines in drinking water. *Green Chemistry Letter and Reviews,* 2014 7, 271-277.

Benvenuto, F; Marín, JM; Sancho, JV; Canobbio, S; Mezzanotte, V; Hernández, F. Simultaneous determination of triazines and their main transformation products in surface and urban wastewater by ultra-high-pressure liquid chromatography-tandem mass spectrometry. *Analytical and Bioanalytical Chemistry,* 2010 397, 2791-2805.

Berzas-Nevado, JJ; Guiberteau-Cabanillas, C; Villaseñor-Llerena, MJ; Rodríguez-Robledo, V. Sensitive SPE GC-MS-SIM screening of endocrine-disrupting herbicides and related degradation products in natural surface waters and robustness study. *Microchemical Journal,* 2007 87, 62-71.

Bolzan, CM; Caldas SS; Guimaraes BS; Primel EG. Dispersive liquid-liquid microextraction with liquid chromatography-tandem mass spectrometry for the determination of triazine, neonicotinoid, triazole and imidazolinone pesticides in mineral water samples. *Journal of the Brazilian Chemical Society,* 2015 26, 1902-1913.

Bottoni, P; Grenni, P; Lucentini, L; Caracciolo, AB. Terbuthylazine and other triazines in Italian water resources. *Microchemical Journal,* 2013 107, 136-142.

Brix, R; Bahi, N; López de Alda, MJ; Farré, M; Fernández, JM; Barceló, D. Identification of disinfection by-products of selected triazines in drinking water by LC-Q-ToF-MS/MS and evaluation of their toxicity. *Journal of Mass Spectrometry,* 2009 44, 330-337.

Cao, W; Yang, B; Qi, F; Qian, L; Li, J; Lu, L; Xu, Q. Simple and sensitive determination of atrazine and its toxic metabolites in environmental water by carboxyl modified polyacrylonitrile nanofibers mat-based solid-phase extraction coupled with liquid chromatography-diode array detection. *Journal of Chromatography A,* 2017 1491, 16-26.

Carabias-Martínez, R; Rodríguez-Gonzalo, E; Fernández-Laespada, ME; Sánchez-San Roman, FJ. Evaluation of surface- and ground-water pollution due to herbicides in agricultural areas of Zamora and Salamanca (Spain). *Journal of Chromatography A,* 2000 869, 471-480.

Carabias-Martínez, R; Rodríguez-Gonzalo, E; Herrero Hernández, E; Sánchez-San Román, FJ; Prado Flores, MG. Determination of herbicides and metabolites by solid-phase extraction and liquid chromatography Evaluation of pollution due to herbicides in surface and groundwaters. *Journal of Chromatography A,* 2002 950, 157-166.

Carabias-Martínez, R; Rodríguez-Gonzalo, E; Fernández-Laespada, ME; Calvo-Seronero, L; Sánchez-San Román, FJ. Evolution over time of the agricultural pollution of waters in an area of Salamanca and Zamora (Spain). *Water Research,* 2003 37, 928-938.

Carabias-Martínez, R; Rodríguez-Gonzalo, E; Miranda-Cruz, EM; Domínguez-Alvarez, J; Hernández-Méndez, J. Comparison of a non-aqueous capillary electrophoresis method with high performance liquid chromatography for the determination of herbicides and metabolites in water samples. *Journal of Chromatography A,* 2006 1122, 194-201.

Chen, J; Bai, L; Tian, M; Zhou, X; Zhang, Y. Hollow-fiber membrane tube embedded with a molecularly imprinted monolithic bar for the microextraction of triazine pesticides. *Analytical Methods,* 2014 6, 602-608.

Chen, D; Zhang, Y; Miao, H; Zhao, Y; Wu, Y. Determination of triazine herbicides in drinking water by dispersive micro solid phase extraction with ultrahigh-performance liquid chromatography-high-resolution mass spectrometric detection. *Journal of Agricultural and Food Chemistry,* 2015 63, 9855-9862.

Commission Decision 2455/2001/EC implementing Council Directive 2000/60/EC establishing the list of priority substances in the field of water. *Official Journal of European Communities,* 2001 L331, 1-5.

Commission Directive 98/83/EC on the quality of water intended for human consumption. *Official Journal of European Communities,* 1998 L330, 32.

Commission Directive 2000/60/EC Establishment and Framework for Community Action in the Field of Water Policy. *Official Journal of European Communities,* 2000 L327, 1-73.

Commission Directive 2008/105/EC Environmental Quality Standards in the Field of Water Policy. *Official Journal of European Communities,* 2008 L348, 84-97.

Commision Directive 2013/39/EU amending Directives 2000/60/EC and 2008/105/EC Regards Priority Substances in the Field of Water Policy. *Official Journal of European Communities,* 2013 L226, 1-17.

Cotton, J; Leroux, F; Broudin, S; Poirel, M; Corman, B; Junot, C; Ducruix, C. Development and validation of a multiresidue method for the analysis of more than 500 pesticides and drugs in water based on on-line and liquid chromatography coupled to high resolution mass spectrometry. *Water Research,* 2016 104, 20-27.

De Toffoli, A; Fumes, B; Lancas, F. Packed in-tube solid phase microextraction with graphene oxide supported on aminopropyl silica: Determination of target triazines in water samples. *Journal of Environmental Science and Health. Part. B, Pesticides, Food, Contaminants, and Agricultural Wastes.* 2018, doi: 10.1080/03601234. 2018.1438831.

Djozan, D; Ebrahimi, B. Preparation of a new solid phase miscro extraction fiber on the basis of atrazine-molecular imprinted polymer: Application for GC and GC/MS screening of triazine herbicides in water, rice and onion. *Analytica Chimica Acta,* 2008 616, 152-159.

Djozan, D; Ebrahimi, B; Mahkam, M; Farajzadeh, MA. Evaluation of a new method for chemical coating of aluminum wire with molecularly imprinted polymer layer. Application for the fabrication of triazines selective solid-phase microextraction fiber. *Analytica Chimica Acta,* 2010 674, 40-48.

Djozan, D; Farajzadeh, MA; Sorouraddin, SM; Baheri, T; Norouzi, J. Inside-needle extraction method based on molecularly imprinted polymer for solid-phase dynamic extraction and preconcentration of triazine herbicides followed by GC–FID determination. *Chromatographia,* 2012 75, 139-148.

Dujaković, N; Grujić, S; Radisić, M; Vasiljević, T; Lausević, M. Determination of pesticides in surface and ground waters by liquid chromatography-electrospray-tandem mass spectrometry. *Analytica Chimica Acta,* 2010 678, 63-72.

Environmental Protection Agency of United States. Final List of Initial Pesticide Active Ingredients and Pesticide Inert Ingredients to be screened under the Federal food, drug and cosmetic act. *Federal Register,* 2009 74, 17579-17585.

Feizbakhsh, A; Ehteshami, S. Polythiophene-Chitosan Magnetic Nanocomposite as a Novel Sorbent for Disperse Magnetic Solid Phase Extraction of Triazine Herbicides in Aquatic Media. *Chromatographia,* 2016 79, 1177-1185.

Feizbakhsh, A; Ehteshami, S. Modified Magnetic Nanoparticles as a Novel Sorbent for Dispersive Magnetic Solid-Phase Extraction of Triazine Herbicides in Aqueous Media. *Journal of AOAC International,* 2017 100, 198-205.

Fingler, S; Mendas, G; Dvorscak, M; Stipicevic, S; Vasilic, Z; Drevenkar, V. Herbicide micropollutants in surface, ground and drinking waters within and near the area of Zagreb, Croatia. *Environmental Science and Pollution Research,* 2017 24, 11017-11030.

García-Galán, MJ; Díaz-Cruz, MS; Barceló, D. Determination of triazines and their metabolites in environmental samples using molecularly imprinted polymer extraction, pressurized liquid extraction and LC-tandem mass spectrometry. *Journal of Hydrology,* 2010 383, 30-38.

Gervais, G; Brosillon, S; Laplanche, A; Helen, C. Ultra-pressure liquid chromatography-electrospray tándem mass spectrometry for multiresidue determination of pesticides in water. *Journal of Chromatography A,* 2008 1202, 163-172.

Gilart, N; Borrul, F; Fontanals, N; Marcé, RM. Selective materials for solid-phase extraction in environmental. *Trends in Environmental Analytical Chemistry,* 2014 1, 8-18.

González-Castro, MJ; Castro-Bustelo, V; Rodríguez-González, N; Beceiro-González, E. Validation of a matrix solid phase dispersion methodology for the determination of triazines herbicides in fish. In: A. Haynes. *Advances in Food Analysis Research.* USA: Nova Science Publishers, Inc.; 2015; 167-176.

González-Castro, MJ; Rodríguez-González, N; Beceiro-González, E. Optimization of a HPLC-UV method for the analysis of chloro-s-triazines in seawater samples. *Current topics in Analytical Chemistry,* 2016 10, 23-28.

Herrero-Hernández, E; Pose-Juan, E; Álvarez-Martín, A; Andrades, MS; Rodríguez-Cruz, MS; Sánchez-Martín, MJ. Pesticides and degradation products in groundwaters from a vineyard region: Optimization of a multiresidue method based on SPE and GC-MS. *Journal of Separation Science,* 2012 35, 3492-3500.

Herrero-Hernández, E; Rodríguez-Cruz, MS; Pose-Juan, E; Sánchez-González, S; Andrades, MS; Sánchez-Martín, MJ. Seasonal distribution of herbicide and insecticide residues in the water resources of the vineyard region of La Rioja (Spain). *Science of the Total Environment,* 2017a 609, 161-171.

Hildebrandt, A; Guillamón, M; Lacorte, S; Tauler, R; Barceló, D. Impact of pesticides used in agriculture and vineyards to surface and groundwater quality (North Spain). *Water Research,* 2008 42, 3315-3326.

Hu, YL; Wang, YY; Hu, YF; Li, GK. Liquid-liquid-solid microextraction based on membrane-protected molecularly imprinted polymer fiber for trace analysis of triazines in complex aqueous samples. *Journal of Chromatography A,* 2009 1216, 8304-8311.

Huff, TB; Foster, GD. Parts-per-trillion LC-MS (Q) analysis of herbicides and transformation products in surface water. *Journal of Environmental Science and Health Part B,* 2011 46, 723-734.

Hurtado-Sánchez, MC; Romero-González, R; Rodríguez-Cáceres, MI; Durán-Merás, I; Garrido-Frenich, A. Rapid and sensitive on-line solid phase extraction-ultra high performance liquid chromatography-electrospray-tandem mass spectrometry analysis of pesticides in surface waters. *Journal of Chromatography A,* 2013 1305, 193-202.

Jiang, H; Adams, CD; Koffskey, W. Determination of chloro-s-triazines including didealkylatrazine using solid-phase extraction coupled with gas chromatography–mass spectrometry. *Journal of Chromatography A,* 2005 1064, 219-226.

Jiménez-Soto, JM; Cárdenas, S; Valcárcel, M. Dispersive micro solid-phase extraction of triazines from waters using oxidized single-walled carbon nanohorns as sorbent. *Journal of Chromatography A*, 2012 1245, 17-23.

Kalogridi, EC; Christophoridis, C; Bizani, E; Drimaropoulou, G; Fytianos, K. Part I: temporal and spatial distribution of multiclass pesticide residues in lake waters of northern Greece: application of an optimized MAE-LC-MS/MS pretreatment and analytical method. *Environmental Science and Pollution Research,* 2014 21, 7239-7251.

Katsumata, H; Kojima, H; Kaneco, S; Suzuki, T; Ohta, K. Preconcentration of atrazine and simazine with multiwalled carbon nanotubes as solid-phase extraction disk. *Microchemical Journal,* 2010 96, 348-351.

Köck, M; Farré, M; Martínez, E; Gajda-Schrantz, K; Ginebreda, A; Navarro, A. Integrated toxicological and chemical approach for the assessment of pesticide pollution in the Ebro river delta (Spain). *Journal of Hydrology,* 2010 383, 73-82.

Köck-Schulmeyer, M; Ginebreda, A; González, S; Cortina, JL; López de Alda, M; Barceló, D. Analysis of the occurrence and risk assessment of polar pesticides in the Llobregat River Basin (NE Spain). *Chemosphere,* 2012 86, 8-16.

Korosa, A; Auersperger, P; Mali, N. Determination of micro-organic contaminants in groundwater (Maribor, Slovenia). *Science of the Total Environment,* 2016 571, 1419-1431.

Kueseng, P; Noir, ML; Mattiasson, B; Thavarungkul, P; Kanatharana, P. Molecularly imprinted polymer for analysis of trace atrazine herbicide in water. *Journal of Environmental Science and Health Part B*, 2009 44, 772-780.

Li, NY; Wu, HL; Qing, XD; Li, Q; Li, SF; Fu, HY; Yu, YJ; Yu, RQ. Quantitative analysis of triazine herbicides in environmental samples by using high performance liquid chromatography and diode array detection combined with second-order calibration based on an alternating penalty trilinear decomposition algorithm. *Analytica Chimica Acta*, 2010 678, 26-33.

Liska, I. On-line versus off-line solid-phase extraction in the determination of organic contaminants in water. *Journal of Chromatography A,* 1993 655, 163-176.

Lissalde, S; Mazzella, N; Fauvelle, V; Delmas, F; Mazellier, P; Legube, B. Liquid chromatography coupled with tandem mass spectrometry method for thirty-three pesticides in natural water and comparison of performance between classical solid phase extraction and passive sampling approaches. *Journal of Chromatography A,* 2011 1218, 1492-1502.

Loos, R; Locoro, G; Contini, S. Occurrence of polar organic contaminants in the dissolved water phase of the Danube River and its major tributaries using SPE-LC-MS2 analysis. *Water Research,* 2010 44, 2325-2335.

Loos, R; Tavazzi, S; Mariani, G; Suurkuusk, G; Paracchini, B; Umlauf, G. Analysis of emerging organic contaminants in water, fish and suspended particulate matter (SPM) in the Joint Danube Survey using solid-phase extraction followed by UHPLC-MS-MS and GC-MS analysis. *Science of the Total Environment,* 2017 607, 1201-1212.

Masiá, A; Campo, J; Vázquez-Roig, P; Blanco, C; Picó, Y. Screening of currently used pesticides in water, sediments and biota of the Guadalquivir River Basin (Spain). *Journal of Hazardous Materials,* 2013a 263, 95-104.

Masiá, A; Ibáñez, M; Blasco, C; Sancho, JV; Picó, Y; Hernández, F. Combined use of liquid chromatography triple quadrupole mass spectrometry and liquid chromatography quadrupole time-of-flight mass spectrometry in systematic screening of pesticides and other contaminants in water samples. *Analytica Chimica Acta,* 2013b 761, 117-127.

Matamoros, V; Jover, E; Bayona, JM. Part-per-Trillion Determination of Pharmaceuticals, Pesticides, and Related Organic Contaminants in River Water by Solid-Phase Extraction Followed by Comprehensive Two-Dimensional Gas Chromatography Time-of-Flight Mass Spectrometry. *Analytical Chemistry,* 2010 82, 699-706.

Mazzella, N; Delmas, F; Delest, B; Méchin, B; Madigou, C; Allenou, JP; Gabellec, R; Caquet, Th. Investigation of the matrix effects on a HPLC-ESI-MS/MS method and application for monitoring triazine, phenylurea and chloroacetanilide concentrations in fresh and estuarine waters. *Journal of Environmental Monitoring,* 2009 11, 108-115.

Mohammadi, A; Ameli, A; Alizadeh, N. Headspace solid-phase microextraction using a dodecylsulfate-doped polypyrrole film coupled to ion mobility spectrometry for the simultaneous determination of atrazine and ametryn in soil and water samples. *Talanta,* 2009 78, 1107-1114.

Moliner-Martínez, Y; Serra-Mora, P; Verdú-Andrés, J; Herráez-Hernández, R; Campíns-Falcó, P. Analysis of polar triazines and degradation products in waters by in-tube solid-phase microextraction and capillary chromatography: an environmentally friendly method. *Analytical and Bioanalytical Chemistry,* 2015 407, 1485-1497.

Montoro, EP; González, RR; Frenich, AG; Torres, MEH; Vidal, JLM. Fast determination of herbicides in waters by ultra-performance liquid chromatography/tandem mass spectrometry. *Rapid Communications in Mass Spectrometry,* 2007 21, 3585-3592.

Nagaraju, D; Huang, SD. Determination of triazine herbicides in aqueous samples by dispersive liquid-liquid microextraction with gas chromatography-ion trap mass spectrometry. *Journal of Chromatography A,* 2007 1161, 89-97.

Navarro, A; Tauler, R; Lacorte, S; Barceló, D. Occurrence and transport of pesticides and alkylphenols in water samples along the Ebro River Basin. *Journal of Hydrology,* 2010 383, 18-29.

Nasrollahpour, A; Moradi, SE. Simple Vortex-Assisted Magnetic Dispersive Solid Phase Microextraction System for Preconcentration and Separation of Triazine Herbicides from Environmental Water and Vegetable Samples Using Fe_3O_4@MIL-100 (Fe) Sorbent. *Journal of AOAC International,* 2018, doi: 10.5740/jaoacint.17-0374.

Papadopoulos, N; Gikas, E; Zalidis, G; Tsarbopoulos, A. Simultaneous determination of Terbuthylazine and its major hydroxy and dealkylated metabolites in wetland water samples using solid-phase extraction and high-performance liquid chromatography with diode-array detection. *Journal of Agricultural and Food Chemistry,* 2007 55, 7270-7277.

Papadopoulos, NG; Gikas, E; Zalidis, G; Tsarbopoulos, A. Determination of herbicide terbuthylazine and its major hydroxyl and dealkylated metabolites in constructed wetland sediments using solid phase extraction and high performance liquid chromatography-diode array detection. *International Journal of Environmental Analytical Chemistry,* 2012 92, 1429-1442.

Passeport, E; Guenne, A; Culhaoglu, T; Moreau, S; Bouyé, JM; Tournebize, J. Design of experiments and detailed uncertainty analysis to develop and validate a solid-phase microextraction/gas chromato-graphy-mass spectrometry method for the simultaneous analysis of 16 pesticides in water. *Journal of Chromatography A,* 2010 1217, 5317-5327.

Portolés, T; Pitarch, E; Lopez, FJ; Hernandez, F. Development and validation of a rapid and wide-scope qualitative screening method for detection and identification of organic pollutants in natural water and wastewater by gas chromatography time-of-flight mass spectrometry. *Journal of Chromatography A,* 2011 1218, 303-315.

Portugal, FCM; Pinto, ML; Nogueira, JMF. Optimization of Polyurethane foams for enhanced stir bar sorptive extraction of Triazinic herbicides in water matrices. *Talanta,* 2008 77, 765-773.

Postigo, C; López de Alda, MJ; Barceló, D; Ginebreda, A; Garrido, T; Fraile, J. Analysis and occurrence of selected medium to highly polar pesticides in groundwater of Catalonia (NE Spain): An approach based on on-line solid phase extraction-liquid chromatography-electrospray-tandem mass spectrometry detection. *Journal of Hydrology,* 2010 383, 83-92.

Retamal, M; Costa, C; Suárez, JM; Richter, P. Multi-determination of organic pollutants in water by gas chromatography coupled to triple quadrupole mass spectrometry. *International Journal of Environmental Analytical Chemistry,* 2013 93, 93-107.

Rimayi, C; Odusanya, D; Weiss, JM; de Boer, J; Chimuka, L. Seasonal variation of chloro-s-triazines in the Hartbeespoort Dam catchment, South Africa. *Science of the Total Environment,* 2018 613, 472-482.

Rocha, MJ; Ribeiro, MFT; Cruzeiro, C; Figueiredo, F; Rocha, E. Development and validation of a GC-MS method for determination of 39 common pesticides in estuarine water – targeting hazardous amounts in the Douro River estuary. *International Journal of Environmental Analytical Chemistry,* 2012 92, 1587-1608.

Rocha, AA; Monteiro, SH; Andrade, GCRM; Vilca, FZ; Tornisielo, VL. Monitoring of pesticide residues in surface and subsurface waters, sediments and fish in center-pivot irrigation areas. *Journal of the Brazilian Chemical Society,* 2015 26, 2269-2278.

Rodríguez-González, N; Beceiro-González, E; González-Castro, MJ; Muniategui-Lorenzo, S. Application of a developed method for the extraction of triazines in surface waters and storage prior to analysis to seawaters of Galicia (Northwest Spain). *Scientific World Journal,* 2013, doi: 10.1155/2013/536369.

Rodríguez-González, N; González-Castro, MJ; Beceiro-González, E; Muniategui-Lorenzo, S; Prada-Rodríguez, D. Determination of Triazine Herbicides in Seaweeds: Development of a sample preparation method based on Matrix Solid Phase Dispersion and Solid Phase Extraction Clean-up. *Talanta,* 2014 121, 194-198.

Rodríguez-González, N; Beceiro-González, E; González-Castro, MJ; Muniategui-Lorenzo, S. An environmentally friendly method for determination of triazine herb icides in estuarine seawater samples by dispersive liquid-liquid microextraction. *Environmental Science and Pollution Research,* 2015a 22, 618-626.

Rodríguez-González, N; González-Castro, MJ; Beceiro-González, E; Muniategui-Lorenzo, S. Development of a Matrix Solid Phase Dispersion methodology for the determination of triazine herbicides in mussels. *Food Chemistry,* 2015b 173, 391-396.

Rodríguez-González, N; Beceiro-González, E; González-Castro, MJ; Alpendurada, MF. On-line solid-phase extraction method for determination of triazine herbicides and degradation products in seawater by ultra-pressure liquid chromatography-tandem mass spectrometry. *Journal of Chromatography A*, 2016 1470, 33-41.

Rodríguez-González, N; Uzal-Varela, R; González-Castro, MJ; Muniategui-Lorenzo, S; Beceiro-González, E. Reliable methods for determination of triazine herbicides and their degradation products in seawater and marine sediments using liquid chromatography-tandem mass spectrometry. *Environmental Science and Pollution Research*, 2017a 24, 7764-7775.

Rodríguez-González, N; González-Castro, MJ; Beceiro-González, E; Muniategui-Lorenzo, S. Development of a matrix solid phase dispersion methodology for the determination of triazine herbicides in marine sediments. *Microchemical Journal*, 2017b 133, 137-143.

Roldán-Pijuán, M; Lucena, R; Cárdenas, S; Valcárcel, M; Kabir, A; Furton, KG. Stir fabric phase sorptive extraction for the determination of triazineherbicides in environmental waters by liquid chromato-graphy. *Journal of Chromatography A*, 2015 1376, 35-45.

Sambe, H; Hoshina, K; Haginaka, J. Molecularly imprinted polymers for triazine herbicides prepared by multi-step swelling and polymerization method. Their application to the determination of methylthiotriazine herbicides in river water. *Journal of Chromatography A*, 2007 1152, 130-137.

Sanagi, MM; Abbas, HH; Ibrahim, WAW; Aboul-Enien, HY. Dispersive liquid-liquid microextraction method based on solidification of floating organic droplet for the determination of triazine herbicides in water and sugarcane samples. *Food Chemistry*, 2012 133, 557-562.

Sanagi, MM; Muhammad SS; Hussain, I; Ibrahim, WAW; Ali, I. Novel solid-phase membrane tip extraction and gas chromatography with mass spectrometry methods for the rapid analysis of triazine herbicides in real waters. *Journal of Separation Science*, 2015 38, 433-438.

Sánchez-Ortega, A; Unceta, N; Gómez-Caballero, A. Sensitive determination of triazines in underground waters using stir bar sorptive extraction directly coupled to automated thermal desorption and gas chromatography-mass spectrometry. *Analytica Chimica Acta*, 2009 641, 110-116.

Santos, TG; Martínez, CBR. Atrazine promotes biochemical changes and DNA damage in a Neotropical fish species. *Chemosphere*, 2012 89, 1118-1125.

See, HH; Sanagi, MM; Ibrahim, WAW; Naim, AA. Determination of triazine herbicides using membrane-protected carbon nanotubes solid phase membrane tip extraction prior to micro-liquid chromatography. *Journal of Chromatography A*, 2010 1217, 1767-1772.

Serra-Mora, P; Jornet-Martinez, N; Moliner-Martínez, Y; Campíns-Falcó, P. In tube-solid phase microextraction-nano liquid chromatography: Application to the determination of intact and degraded polar triazines in waters and recovered struvite. *Journal of Chromatography A*, 2017 1513, 51-58.

Song, Y; Zhao, S; Tchounwou, P; Liu, YM. A nanoparticle-based solid-phase extraction method for liquid chromatography–electrospray ionization-tandem mass spectrometric analysis. *Journal of Chromatography A*, 2007 1166, 79-84.

Sorouraddin, SM; Mogaddam, MRA. Development of molecularly imprinted-solid phase extraction combined with dispersive liquid-liquid microextraction for selective extraction and preconcentration of triazine herbicides from aqueous samples. *Journal of the Iranian Chemical Society*, 2016 13, 1093-1104.

Terzopoulou, E; Voutsa, D; Kaklamanos, G. A multi-residue method for determination of 70 organic micropollutants in surface waters by solid-phase extraction followed by gas chromatography coupled to tandem mass spectrometry. *Environmental Science and Pollution Research*, 2015 22, 1095-1112.

Trenholm, RA; Vanderford, BJ; Snyder, SA. On-line solid phase extraction LC–MS/MS analysis of pharmaceutical indicators in water: A green alternative to conventional methods. *Talanta*, 2009 79, 1425-1432.

Trić-Petrović, T; Dordević, J; Dujaković, N; Kumrić, K; Vasiljević, T; Laušević, M. Determination of selected pesticides in environmental water by employing liquid-phase microextraction and liquid chromatography-tandem mass spectrometry. *Analytical and Bioanalytical Chemistry*, 2010 397, 2233-2243.

van Pinxteren, M; Bauer, C; Popp, P. High performance liquid chromatography-tandem mass spectrometry for the analysis of 10 pesticides in water: A comparison between membrane-assisted solvent extraction and solid phase extraction. *Journal of Chromatography A,* 2009 1216, 5800-5806.

Wang, SL; Ren, LP; Liu, CY; Ge, J; Liu, FM. Determination of five polar herbicides in water samples by ionic liquid dispersive liquid-phase microextraction. *Analytical and Bioanalytical Chemistry,* 2010 397, 3089-3095.

Wang, C; Ji, S; Wu, Q; Wu, C; Wang, Z. Determination of triazine herbicides in environmental samples by dispersive liquid-liquid microextraction coupled with high performance liquid chromatography. *Journal of Chromatographic Science,* 2011 49, 689-694.

Wang, Y; You, J; Ren, R; Xiao, Y; Gao, S; Zhang, H; Yu, A. Determination of triazines in honey by dispersive liquid-liquid microextraction high-performance liquid chromatography. *Journal of Chromatography A,* 2010 1217, 4241-4246.

Wu, Q; Feng, C; Zhao, G; Wang, C; Wang, Z. Graphene-coated fiber for solid-phase microextraction of triazine herbicides in water samples. *Journal of Separation Science,* 2012 35, 193-199.

Wu, XL; Meng, L; Wu, Y; Luk, YY; Ma, Y; Du, Y. Evaluation of Graphene for Dispersive Solid-Phase Extraction of Triazine and Neonicotine Pesticides from Environmental Water. *Journal of the Brazilian Chemical Society,* 2015 26, 131-139.

Yang, BY; Qi, FF; Li, XQ; Liu, JJ; Rong, F; Xu, Q. Application of Nylon6/Polypyrrole core–shell nanofibres mat as solid-phase extraction adsorbent for the determination of atrazine in environmental water samples. *International Journal of Environmental Analytical Chemistry,* 2015 95, 1112-1123.

Yang, Q; Chen, B; He, M; Hu, B. Sensitive determination of seven triazine herbicide in honey, tomato and environmental water samples by hollow fiber based liquid-liquid-liquid microextraction combined with sweeping micellar electrokinetic capillary chromatography. *Talanta,* 2018 186, 88-96.

Ye, C; Zhou, Q; Wang, X. Improved single-drop microextraction for high sensitive analysis. *Journal of Chromatography A,* 2007 1139, 7-13.

Zhang, G; Zang, X; Zhou, X; Wang, L, Wang, C; Wang, Z. Extraction of triazine herbicides from environmental water samples with magnetic graphene nanoparticles as the adsorbent followed by determination using gas chromatography-mass spectrometry. *Chinese Journal of Chromatography,* 2013 31, 1071-1075.

Zhang, P; Bui, A; Rose, G; Allinson, G. Mixed-mode solid-phase extraction coupled with liquid chromatography tandem mass spectrometry to determine phenoxy acid, sulfonylurea, triazine and other selected herbicides at nanogram per litre levels in environmental waters. *Journal of Chromatography A,* 2014 1325, 56-64.

Zhang, Z; Mei, M; Huang, Y; Huang, X; Huang, H; Ding, Y. Facile preparation of a polydopamine-based monolith for multiple monolithic fiber solid-phase microextraction of triazine herbicides in environmental water samples. *Journal of Separation Science*, 2017 40, 733-743.

Zhang, H; Yuan, Y; Sun, Y; Niu, C; Qiao, F; Yan, H. An ionic liquid-magnetic graphene composite for magnet dispersive solid-phase extraction of triazine herbicides in surface water followed by high performance liquid chromatography. *The Analyst,* 2018 143, 175-181.

Zhao, RS; Yuang, JP; Jiang, T; Shi, JB; Cheng, CG. Application of bamboo charcoal as solid-phase extraction adsorbent for the determination of atrazine and simazine in environmental water samples by high-performance liquid chromatography-ultraviolet detector. *Talanta,* 2008 76, 956-959.

Zhao, G; Song, S; Wang, C, Wu, Q; Wang, Z. Determination of triazine herbicides in environmental water samples by high-performance liquid chromatography using graphene-coated magnetic nanoparticles as adsorbent. *Analytica Chimica Acta,* 2011 708, 155-159.

Zhou, Q; Xiao, J; Wang, W; Liu, G; Shi, Q; Wang, J. Determination of atrazine and simazine in environmental water samples using multiwalled carbon nanotubes as the adsorbents for preconcentration prior to high performance liquid chromatography with diode array detector. *Talanta*, 2006 68, 1309-1315.

Zhou, Q; Pang, L; Xie, GH; Xiao, JP; Bai, HH. Determination of atrazine and simazine in environmental water samples by dispersive liquid-liquid microextraction with high performance liquid chromatography. *Analytical Science*, 2009 25, 73-76.

Zhou, L; Su, P; Deng, YL; Yang, Y. Self-assembled magnetic nanoparticle supported zeolitic imidazolate framework-8: An efficient adsorbent for the enrichment of triazine herbicides from fruit, vegetables, and water. *Journal of Separation Science*, 2017 40, 909-918.

Zhou, T; Ding, J; He, Z; Li, J; Liang, Z; Li, C; Li, Y; Chen, Y; Ding, L. Preparation of magnetic superhydrophilic molecularly imprinted composite resin based on multi-walled carbon nanotubes to detect triazines in environmental water. *Chemical Engineering Journal*, 2018 334, 2293-2302.

In: Solid-Phase Extraction
Editor: Ben Benson

ISBN: 978-1-53614-582-3
© 2019 Nova Science Publishers, Inc.

Chapter 3

DETERMINATION OF TRACE ELEMENTS IN WATER USING A SOLID-PHASE EXTRACTION DISK

*Kenta Hagiwara**, *PhD and Yuya Koike, PhD*
Department of Applied Chemistry,
Meiji University, Kawasaki, Kanagawa, Japan

ABSTRACT

Solid-phase extraction (SPE) has been widely used in the pharmaceutical, industrial, and environmental fields to separate and/or preconcentrate inorganic and organic analytes. In particular, membrane disk SPEs have been used to pretreat large-volume water samples because their large cross-section areas allow relatively high flow rates. SPE disks maintain their shape from the pretreatment process to the assaying step, making them useful for wet analyses as well as direct analyses, e.g., X-ray fluorescence spectrometry and γ-ray spectrometry. In addition, miniature SPE disks can be used in combination with portable analyzers for on-site analysis. In this chapter, rapid and simple methods combining

---

* Corresponding Author Email: kenhagi@meiji.ac.jp.

disk SPE with several detection techniques for the determination of trace elements in water are described.

Keywords: solid-phase extraction disk, atomic spectrometry, X-ray fluorescence spectrometry, γ-ray spectrometry, water

1. INTRODUCTION

Solid-phase extraction (SPE) has been widely used in the pharmaceutical, industrial, and environmental fields to separate and/or preconcentrate inorganic and organic analytes. At present, various types of SPE agents are commercially available; the user must select the optimum SPE agent to accurately determine a given analyte. The classes of SPE sorbents include iminodiacetate chelating resins, ion exchange resins, hydrophobic resins, activated carbon beads, and macrocyclic molecular recognition resins. The base, function, particle size, and filler content of the SPE sorbent all have an influence on the adsorption and separation performance. Additionally, the shape of the SPE formats has a bearing on the SPE handling procedure and the subsequent detection technique.

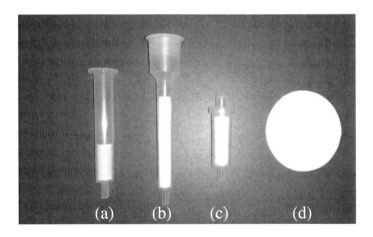

Figure 1. Formats for solid-phase extraction. (a) Regular syringe barrel, (b) Large volume syringe barrel, (c) cartridge, (d) disk.

Figure 1 shows the differences among the four types of SPE formats. SPE disks differ from SPE cartridges or syringes in that the disk is a membrane loaded with a solid sorbent, whereas the cartridge or syringe contains the sorbent. For example, Empore disk (3M) is produced by trapping microscopic sorbent beads within an inert matrix of polytetrafluoroethylene (PTFE) fibers. The ENVI-Disk (Sigma Aldrich) has a porous glass fiber matrix containing surface-modified silica. In general, commercially available SPE disks have a thickness of 0.5 mm and a diameters of 47 mm, and are used with a dedicated manifold (Figure 2) and a vacuum pump for sample and eluent flow. However, SPE disks can be cut using an edged tool, so SPE disks of an arbitrary diameter can be prepared. These miniature disks can be used in a plastic syringe with a filter holder to provide the sample flow, and the sample solution can be passed easily through the SPE disk by pressurization, as with a SPE cartridge or syringe. Miniature disks can also be added to a flow injection system for on-line SPE preconcentration.

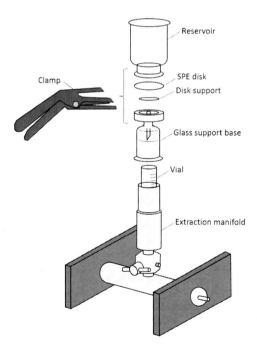

Figure 2. Manifold system for SPE disk.

SPE disks have been used to pretreat large volume water samples because their large cross-sectional areas allow relatively high flow rates. SPE disks maintain their shape from the pretreatment process to the assaying step, so they are useful for wet analyses as well as direct analyses, e.g., X-ray fluorescence (XRF) spectrometry and γ-ray spectrometry. In this chapter, rapid and simple methods combining disk SPE with several detection techniques for the determination of trace elements in water are described.

2. ATOMIC SPECTROMETRY

2.1. Determination of Trace Multivalent Transition Elements

When analyzing the elements in an aqueous sample by an atomic spectrometry technique such as inductively coupled plasma-atomic emission spectrometry (ICP-AES), inductively coupled plasma-mass spectrometry (ICP-MS), or atomic absorption spectrometry (AAS), a SPE procedure involving the following steps is carried out: sample preparation (measuring the sample volume and pH adjustment), SPE disk conditioning, sample extraction, washing of the SPE disk, sample elution, and dilution of the eluent to the desired volume. The procedures when using a SPE disk are the same as those carried out when using a SPE cartridge or syringe.

The analytical procedure for the simultaneous determination of Al, Mn, Fe, Co, Ni, Cu, Zn, Cd, and Pb via concentration with an iminodiacetate chelating disk (ICD) and ICP-AES [1] are shown in Figure 3. A 1000 mL sample of environmental water from which the suspended matter had been removed was adjusted to pH 5.6, and then passed through an ICD at a flow rate of 150 mL min^{-1} to collect these nine metals. The ICD was washed with 20 mL of pure water. The collected Al, Mn, Fe, Co, Ni, Cu, Zn, Cd, and Pb were eluted with 9 mL of 3 mol L^{-1} nitric acid. The eluate was diluted to 10 mL with pure water, and then analyzed using ICP-AES. The detection limits were 0.003-0.25 μg L^{-1} at an enrichment factor of 100. Spike tests using 0.1 mg L^{-1} of the nine metals in tap water, river water, irrigation water, seawater, industrial wastewater, and treatment wastewater demonstrated that quantitative recoveries were achieved, except in the case of Mn (59%) in the seawater sample.

The ICD/ICP-AES method can be applied for chemical state analysis of metals in water [2]. The sample water was treated using the above method and then underwent acid digestion using nitric acid and perchloric acid. The Al, Mn, Fe, Co, Ni, Cu, Zn, Cd, and Pb in the water were classified as ICD-retained soluble metals, inert soluble metals, and particulate metals. This method was applied to investigate the distribution of metal species in industrial seawater and industrial wastewater. In the seawater samples, 80-100% of the Al, Fe, Co, and Pb were present as particulate metals, and 70-90% of the Cd was present as an ICD-retained soluble metal. The concentration and percentage of the nine metals in each chemical state were different for each of the samples. These results suggested that acid digestion is required for samples containing natural organic chelators when the determination of total soluble metals is necessary.

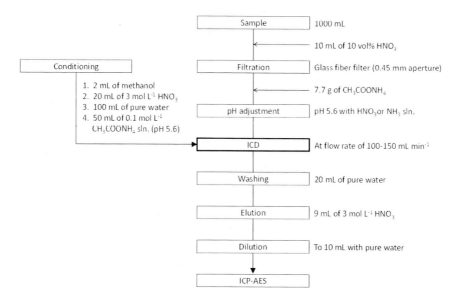

Figure 3. Flowchart of the ICD preconcentration for determination of Al, Mn, Fe, Co, Ni, Cu, Zn, Cd, and Pb by ICP-AES.

The ICD/ICP-AES method can be also applied to the determination of U, which is a radioactive nuclide present in water [3]. An environmental water sample was adjusted to pH 5, and 1 mmol L^{-1} of cyclohexane-diaminetetraacetic acid (CDTA) was added. In this method, CDTA acted as a masking reagent for heavy metals other than U. The solution was passed through an ICD, and the collected U was then eluted with 7.5 mL of 2 mol L^{-1} nitric acid. The eluate was analyzed by ICP-AES after dilution. A preconcentration factor of 200 was achieved using a 2000 mL sample, and the method was finished in 20 min. The detection limit was 0.05 μg L^{-1}. This method has been applied to the determination of U in seawater and mineral drinking water samples, and the concentrations of U determined were in relatively good agreement with the results obtained using ICP-MS and α-ray spectrometry.

Rare earth elements are present in general environmental at sub-ng L^{-1} to ng L^{-1} concentrations; a combination of ICD preconcentration and ICP-MS [4] is suitable for the determination of these analytes. A 50 mL sample of seawater was adjusted to pH 5.5, and then passed through a miniature ICD (5 mm diameter), to collect Y, La, Ce, Pr, Nd, Sm, Eu, Gd, Tb, Dy, Ho, Er, Tm, Yb, and Lu. The miniature ICD was used to achieve high-enrichment preconcentration of the rare earth elements and to eliminate the alkali and alkaline earth metals in the sample solutions. The collected rare earth elements were eluted by ultrasonication of the disk with 0.5-1 mL of 0.1 mol L^{-1} nitric acid, and then analyzed using ICP-MS. The detection limits of the rare earth elements were 0.03-4.5 ng L^{-1} at enrichment factors of 100.

Miniature disk can be added to a flow injection system for on-line SPE preconcentration [5]. In one example, 1 mL of a sample solution in a sample loop was injected into a stream of a 50 mmol L^{-1} ammonium acetate carrier solution, and then passed through a miniature ICD (5 mm diameter). The heavy metals absorbed onto the disk, in this case, V, Mn, Co, Ni, Cu, Mo, Cd, and U were eluted with 0.5 mL of 0.1 mol L^{-1} HNO_3 and injected into the ICP-MS nebulizer. The processing time for one sample was 350 s, and the detection limits of the heavy metals were 0.006-0.40 μg L^{-1}. The proposed method was evaluated by analyzing certified reference materials of seawater, and the analytical data for the eight heavy metals obtained from the method agreed well with the certified values. This method is applicable for the routine analysis of heavy metals in water.

The elution step may give rise to negative error owing to the incomplete elution of the analytes from the SPE reagent. Direct analysis of the analytes in the SPE reagent is an effective way to avoid analyte loss. For example, the direct application of the AAS method can be employed for the analysis of organic and inorganic materials, and combination with SPE preconcentration [6] is also possible. After pH adjustment (pH 5.6), a 100 mL water sample was passed through an ICD (47 mm diameter) to preconcentrate Co, Ni, Cu, Cd, Sn, Pb, and Bi. The ICD was then dried in an electric oven, and 110-145 small disks (2 mm diameter) were punched out from the ICD. The seven analytes could be determined without an elution step by directly introducing the small disks into a graphite furnace. The concentrations of the analytes were calculated using calibration curves constructed using aqueous standards. When an enrichment factor of 140 was obtained using a 100 mL water sample, the detection limits were 0.077-0.80 µg L^{-1}. A spike test for the seven analytes in tap water, rainwater, river water, and mineral drinking water showed quantitative recoveries (93%-108%).

2.2. Speciation of Toxic Elements

Heavy metals such as As, Se, and Cr are found in water derived from various environmental sources, and are present in multiple chemical forms. For example, ground water contains mainly inorganic arsenite (As(III)) and/or arsenate (As(V)) ions, which are dissolved from soils and minerals. Although fewer cases have been reported, organic As compounds, such as phenylarsonic acid (PAA) and diphenylarsinic acid (DPAA), also appear in drinking water due to human activity [7]. The toxicity levels of heavy metals depend on their chemical form due to differences in their mechanisms of action. Thus, the speciation of heavy metals is required when evaluating the toxicity of water samples. Speciation is also required to select a method for removing the pollutant from water, after pollution is detected.

A speciation procedure for As using three SPE disks and graphite furnace AAS [8] is shown in Figure 4. Here, DPAA, PAA, and inorganic arsenic (iAs) in drinking water were determined. These compounds were collected simultaneously using three stacked SPE disks, i.e., an SDB-XD disk (the upper layer), an activated carbon disk (the middle layer), and a cation exchange disk (CED) loaded with Zr and Ca (ZrCa-CED; the lower layer). In an aqueous solution, these As species exist as anions, and thus were separated and collected by π-π interactions, polar interactions, and complex forming reactions. A 200 mL aqueous sample was adjusted to pH 3, and passed through the SPE disks at a flow rate of 15 mL min^{-1} to concentrate DPAA on the SDB-XD disk, PAA on the activated carbon disk, and iAs on the ZrCa-CED. The As compounds were eluted from the disks with 10 mL of ethanol containing 0.5 mol L^{-1} ammonia solution for DPAA, 20 mL of 1 mol L^{-1} ammonia solution for PAA, and 20 mL of 6 mol L^{-1} hydrochloric acid for iAs. The eluates of DPAA, PAA, and iAs were diluted to 20, 25, and 25 mL with deionized water, respectively, and then analyzed using graphite furnace AAS. The detection limits of As were 0.13 µg L^{-1} and 0.16 µg L^{-1} at enrichment factors of 10 and 8, respectively. Spike tests with 2 µg of DPAA, PAA, and iAs in 200 mL of tap water and mineral drinking water showed good recoveries (96.1-103.8%).

Cr(III) is an essential trace element in the human body, while Cr(VI) exhibits high toxicity and carcinogenicity. Therefore, an environmental quality standard and an emission standard for Cr(VI) have been established in Japan. The following analytical procedure using ion exchange disks and metal furnace AAS was employed for the speciation of trace inorganic Cr in water [9]. A 500 mL water sample was adjusted to pH 3, and then passed through an ICD placed on a CED at a flow rate of 40 mL min^{-1} to collect Cr(III). The filtrate was adjusted to pH 10, and then passed through an anion exchange disk (AED) at a flow rate of 2 mL min^{-1} to collect Cr(VI). The collected Cr(III) and Cr(VI) were eluted with 40 mL of 3 mol L^{-1} nitric acid and 40 mL of 1 g L^{-1} diphenylcarbazide solution, respectively. Each eluate was diluted to 50 mL with deionized water, and then analyzed using metal furnace AAS. The detection limits were 0.08 µg L^{-1} for both Cr(III) and Cr(VI). The recovery tests for 1 µg L^{-1} of Cr(III) and Cr(VI) in tap water, mineral drinking water, hot spring water, and river water samples showed sufficient values (98.1-106%), except in the case of river water sampled downstream, due to the relatively higher chemical oxygen demand value.

A method based on the above technique was developed for the speciation of inorganic and organic Cr [10]. A water sample was adjusted to pH 5.6, and passed through a CED placed on an AED. Cr(III) acetyl acetonate (Cr(acac)$_3$) and Cr(III) were adsorbed on the CED, and Cr(VI) was adsorbed on the AED. The adsorbed Cr(acac)$_3$ was eluted with 50 mL of carbon tetrachloride, followed by the elution of Cr(III) with 50 mL of 3 mol L^{-1} nitric acid. Cr(VI) was eluted with 50 mL of 3 mol L^{-1} nitric acid. All of the eluates were subsequently analyzed using metal furnace AAS. The recovery tests for Cr(III), Cr(VI), and Cr(acac)$_3$ exhibited good results (96.0-107%) when 5 µg of each species was added to 100 mL samples of tap water, rainwater, and mineral drinking water.

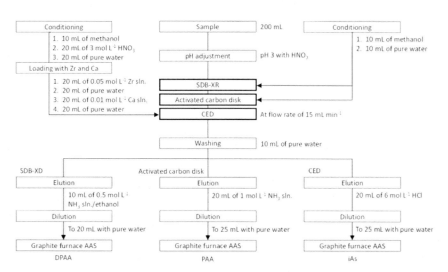

Figure 4. Flowchart of disk SPE preconcentration and separation for determination of DPAA, PAA, and iAs by graphite furnace AAS.

3. X-RAY FLUORESCENCE SPECTROMETRY

3.1. In Lab SPE

SPE/atomic spectrometry methods are simple and useful for sensitive analysis, although such methods are difficult to apply to the screening of a large number of water samples because the elution step is tedious. In contrast, direct analysis of SPE disks using XRF spectrometry does not require an elution step, so it is useful for the screening of trace elements in water.

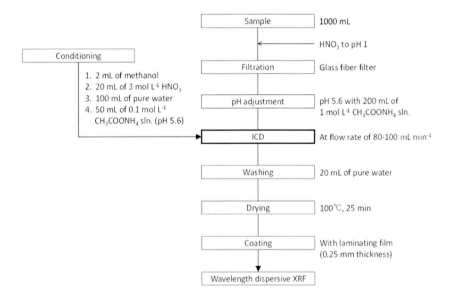

Figure 5. Flowchart of the ICD preconcentration for screening of Mn, Fe, Co, Ni, Cu, Zn, Cd, Hg, and Pb by wavelength dispersive XRF spectrometry.

A screening procedure for Mn, Fe, Co, Ni, Cu, Zn, Cd, Hg and Pb in water using ICD concentration and wavelength dispersive XRF spectrometry [11] is shown in Figure 5. A 1000 mL sample of environmental water was adjusted to pH 5.6, and then passed through an ICD. The ICD was dried in an electric oven, and then coated on both sides with commercially available laminate film to prevent X-ray damage. The disk was analyzed directly using wavelength dispersive XRF spectrometry. The detection limits were 0.1-7 μg L^{-1}. A spike test using 10 μg of analytes in city water, rainwater, river water, lake water, seawater, and hot spring water showed good recoveries (90-100%) except in the case of Mn in city water and rainwater. The analytical results for municipal tap water and rainwater agreed well with the values obtained using AAS. The ICD/atomic spectrometry method is not suitable for determining Hg, because it is extremely difficult to elute Hg adsorbed on ICD from the sorbent. The ICD/XRF method is useful for the determination of Hg in water.

The method using an SPE disk and XRF spectrometry can be applied to the speciation of Cr in water [12]. A water sample was adjusted to pH 3, and then passed through an AED placed on a CED to separate Cr(III) and Cr(VI). These disks were dried and coated with a laminate film. The disks were analyzed using a wavelength dispersive XRF spectrometer. When a 1000 mL water sample was used, the detection limits for Cr(III) and Cr(VI) were 0.17 μg L^{-1} and 0.16 μg L^{-1}, respectively. A spike test using 50 μg L^{-1} of Cr(III) and Cr(VI) in tap water and river water showed quantitative recoveries (94-114%), although this was not observed for mineral drinking water owing to the overlap of V Kβ with Cr Kα. The recovery after overlap correction was satisfactory (115%).

3.2. On-Site SPE

When analytical methods involving chromatographic or extraction separation and atomic spectrometry are used for in-lab speciation of inorganic As or Se in water, these methods must incorporate a rapid assay to obtain the correct quantitative values because inorganic As and Se species tend to be unstable in natural water due to the water chemistry. Some pretreatment techniques to improve the stability of these species have been reported. However, these methods require extreme thermal management, the addition of preservatives, and the rapid removal of suspended matter. These complications can be avoided by the on-site separation of analytes in water. SPE is suitable for on-site separation and collection of analytes in environmental water because it does not require a power supply, numerous acid-base solutions, or organic solvents in the sample flow step. In addition, if an SPE disk is used, direct analysis using XRF spectrometry can be used for the screening and speciation of As and Se in water without an elution step.

The following procedure was used for the speciation of inorganic As using wavelength dispersive XRF spectrometry after on-site disk SPE [13]. In the on-site SPE method, the sample water was passed through a miniature SPE disk (13 mm diameter) by pressurization using a plastic syringe with a filter holder. The removal of Pb^{2+} from the sample water was first conducted to avoid the overlapping Pb Lα (10.55 keV) and As Kα (10.54 keV) signals in the XRF spectrum. To this end, a 50 mL aqueous sample (pH 5-9) was passed through an ICD. The filtrate was adjusted to pH 2-3, and then ammonium pyrrolidine dithiocarbamate (APDC) solution was added. The solution was passed through a hydrophilic PTFE filter placed on a ZrCa-CED at a flow rate of 12.5 mL min^{-1} to separate the As(III)-PDC complex and As(V). Each SPE disk was affixed to an acrylic plate using adhesive cellophane tape, and then examined by wavelength dispersive XRF spectrometry. The detection limits of As(III) and As(V) were 0.8 μg L^{-1} and 0.6 μg L^{-1}, respectively. The proposed method was successfully applied to perform As speciation and concentration evaluation in spring water and well water.

This method can be also applied to the speciation of inorganic Se in water. [14] A 50 mL aqueous sample was adjusted to pH 3, and APDC solution was added. The solution was passed through a PTFE filter placed on an AED to separate the Se(IV)-PDC complex and Se(VI). Each solid-phase extraction disk was affixed to an acrylic plate, and then analyzed by wavelength dispersive XRF spectrometry. This method was successfully used for the speciation of Se in drinking water.

3.3. On-Site Analysis

Handheld XRF spectrometers, which are also referred to as portable field XRF spectrometers, are rechargeable portable element analyzers commonly used for the on-site screening of solid samples. If a handheld XRF spectrometer could evaluate the concentrations of chemical substances in environmental water, the water quality could be investigated during a simultaneous survey of the surrounding environment, such as soil, minerals, and waste, in the field. This would aid in the rapid identification and removal of pollutant sources. The simplest way to detect elements in water by XRF spectrometry is using an aqueous sample holder or a drip filter paper. However, the sensitivities of these methods using a handheld XRF spectrometer are somewhat low, with the detection limits of elements at the mg L^{-1} level [15]. The high concentration of the water sample using a SPE disk is useful for on-site handheld XRF spectrometry.

An analytical method combining a miniature SPE disk (13 mm diameter) with handheld XRF spectrometry for the determination of micrograms per liter concentrations of As [16] is shown in Figure 6. A 50 mL aqueous sample was adjusted to pH 3, and then passed through a Ti- and Zr-loaded activated carbon disk (TiZr-CD) to preconcentrate the As. The SPE disk was affixed to an acrylic plate with cellophane tape, and then examined by handheld XRF spectrometry. The TiZr-CD adsorbed inorganic As (as As(III) and As(V)) and organic As (as methyl, phenyl, and aromatic As compounds) from the water. The limit of detection for As was 2.0 μg L^{-1}. The concentrations of As in well water samples were determined using this method, and the results were similar to those obtained from AAS. This method did not require a power supply, toxic solution, or gas at any analytical step, and therefore it is suitable for the on-site determination of As in environmental water.

This method can be applied to the simultaneous determination of Mn, Fe, Cu, Zn, Cd, Hg, and Pb in drinking water [17] using a miniature ICD (13 mm diameter). The Cd Kα peak, which has a high fluorescence X-ray energy (23.17 keV), of 10 μg Cd adsorbed on the ICD was unclear because of the high background and noise resulting from the scattered X-rays. In order to solve this problem, a metal plate was placed under the ICD. A W plate was most effective for the improvement of the peak to background ratio. Therefore, when Cd was determined, the W plate was placed under the ICD. The detection limits of the heavy metals were 3.2-9.4 μg L⁻¹. These detection limits were comparable with those obtained using ICP-AES. This method can also be applied to the simultaneous determination of As, Se, and Cr(VI) in drinking water [18] using a miniature Ti loaded AED (Ti-AED). By combining a Ti-AED with a ICD, the disk SPE/handheld XRF method can be applied to the on-site determination of ten heavy metals (Cr(VI), Mn Fe, Cu, Zn, As, Se, Cd, Hg, and Pb), which were chosen based on the quality standards for drinking water in Japan, on a μg L⁻¹ scale.

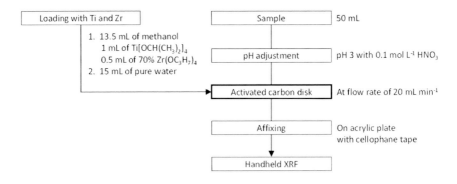

Figure 6. Flowchart of the TiZr-CD preconcentration for on-site screening of As by handheld XRF spectrometry.

4. γ-RAY SPECTROMETRY

The high preconcentration of analytes is necessary to determine trace radioactive nuclides in environmental water using γ-ray spectrometry. The preconcentration of the radioactive nuclides is generally carried out by coprecipitation or evaporation drying methods using several thousands or millions mL of sample water. However, these methods are slow and tedious. On the other hand, preconcentration using SPE disks is simple, and is particularly useful for radioactivity analysis because it allows to rapid preconcentration of large sample volumes, direct analysis by γ-ray spectrometry without an elution step, and sensitive detection without γ-ray self-absorption [19]. The γ-ray spectrometry of radioactive Cs ([134]Cs and ^{137}Cs) in environmental water using a Cs rad disk was reported after the nuclear accident at the Fukushima No. 1 nuclear power plant [20]. A Cs rad disk contains microscopic sorbent beads comprising hexacyanoferrate, which are capable of efficiently and selectively collecting Cs in water. A water sample was passed through a Cs rad disk to preconcentrate radioactive Cs, and then analyzed using a high purity-Ge semiconductor detector. The conditions for quantitative Cs correction were as follows; (a) an adsorption amount of Cs of less than 10 mg per disk, (b) a sample flow rate of less than 400 mL min^{-1}, (c) a sample pH of 2-9, and (d) an amount of K in the sample that results in less than 125 mg of K per disk.

An analytical method using ion exchange disks and γ-ray spectrometry can be applied to determine trace radioactive nuclides having a short half-life in water [21, 22]. By using an ICD for preconcentration, the radioactivity of ^{210}Pb in rain water could be determined by the 46.5 keV γ-ray signal. In addition, short-lived [212]Pb (half-life: 10.6 h) and cosmogenicaly produced [7]Be (half-life: 53.3 days) were detected in rainwater samples. By using an AED placed on a CED for preconcentration, many more radioactive nuclides could be detected. Five natural ([7]Be, [212]Pb, [214]Pb, [212]Bi, and [214]Bi) and three artificial (^{131}I, [134]Cs, and ^{137}Cs) radioactive nuclides were identified in the γ-ray spectra of rainwater sampled at Kawasaki, Japan in 2011. This method could determine extremely short-lived radioactive nuclides such as [214]Pb (half-life: 26.8 min) and [214]Bi (half-life: 19.9 min).

In general, radioactivity analysis is employed in a radiation controlled area, because the estimation of the detection efficiencies for γ-ray spectrometry uses standard radiation sources. However, the detection efficiencies for γ-ray spectrometry can be estimated using the natural radioactive nuclides present in commercially available chemical reagents such as KCl, LaF$_3$, and Lu$_2$O$_3$, which contain [40]K, [138]La, and [176]Lu, respectively [23]. This method enables the determination of radioactive nuclides by γ-ray spectrometry outside of radiation controlled areas. A disk type γ-ray standard source can be also prepared by dropping a KCl or Lu_2O_3 solutions and a gelatin solution on a glass fiber membrane [24]. The combination of this method and disk SPE preconcentration enables the estimation of the activities of radioactive nuclides in environmental water samples regardless of whether the analysis is carried out in a radiation controlled area.

REFERENCES

[1] Kuriyama, K., Ouyang, T., Wang, N., Furusho, Y. (1998). Rapid Extraction of Heavy Metals from Water Using Iminodiacetate Chelating Membrane. *Journal of Industrial Water*, 481: 29-36.

[2] Ouyang, T., Wang, N., Iwashima, K., Kuriyama, K., Furusho, Y. (1999). Simultaneous Multi-element Preconcentration in Aqueous Environmental Samples using Iminodiacetate Extraction Disk (IED) followed by ICO-AES Analysis. *Journal of Environmental Chemistry*, 9: 347-357.

[3] Miura, T., Morimoto, T., Hayano, K., Kishimoto, T. (2000). Determination of uranium in water samples by ICP-AES with chelating resin disk preconcentration. *Bunseki Kagaku*, 49: 245-249.

[4] Oshima, M., Lee, K. H., Gao, Y., Motomizu, O. (2000). Mini-Scale Solid-Phase Collection and Concentration of Ultra-Trace Elements in Natural Waters for the Simultaneous Multielement Determination by Air-Flow Injection/ICP–Mass Spectrometry. *Chemistry Letters*, 29: 1338-1339.

[5] Lee, K. H., Oshima, M., Morimizu, O. (2002). Inductively coupled plasma mass spectrometric determination of heavy metals in sea-water samples after pre-treatment with a chelating resin disk by an on-line flow injection method. *Analyst*, 127: 769-774.

[6] Inui, T., Kosuge, A., Ohbuchi, A., Fujita, K., Koike, Y., Kitano, M., Nakamura, T. (2012). Determination of Heavy Metals at Sub-ppb

Levels in Water by Graphite Furnace Atomic Absorption Spectrometry Using a Direct Introduction Technique after Preconcentration with an Iminodiacetate Extraction Disk. *American Journal of Analytical Chemistry*, 3: 683-692.

[7]   Ishizaki, M., Yanaoka, T., Nakamura, M., Hakuta, T., Ueno, S., Komuro, M., Shibata, M., Kitamura, T., Honda, A., Doy, M., Ishii, K., Tamaoka, A., Shimojo, N., Ogata, T., Nagasawa, E., Hanaoka, S. (2005). Determination of Bis(diphenylarsine)oxide, Diphenylarsinic Acid and Phenylarsonic Acid, Compounds Probably Derived from Chemical Warfare Agents, in Drinking Well Water. *Journal of Health Science*, 51: 130-137.

[8]   Hagiwara, K., Inui, T., Koike, Y., Nakamura, T. (2013). Determination of Diphenylarsinic Acid, Phenylarsonic Acid and Inorganic Arsenic in Drinking Water by Graphite-furnace Atomic-absorption Spectrometry after Simultaneous Separation and Preconcentration with Solid-phase Extraction Disks. *Analytical Sciences*, 29: 1153-1158.

[9]   Inui, T., Fujita, K., Kitano, M., Nakamura, T. (2010). Determination of Cr(III) and Cr(VI) at Sub-ppb Levels in Water with Solid-Phase Extraction/Metal Furnace Atomic Absorption Spectrometry. *Analytical Sciences*, 26: 1093-1098.

[10] Kamakura, N., Iui, T., Kitano, M., Nakamura, T. (2014). Determination of Chromium(III), Chromium(VI), and Chromium(III) acetylacetonate in water by ion-exchange disk extraction/metal furnace atomic absorption spectrometry. *Spectrochimica Acta Part B*, 93: 28-33.

[11] Abe, W., Isaka, S., Koike, Y., Nakano, K., Fujita, K., Nakamura, T. (2006). X-ray fluorescence analysis of trace metals in environmental water using preconcentration with an iminodiacetate extraction disk. *X-ray Spectrometry*, 35: 184-189.

[12] Inui, T., Abe, W., Kitano, M., Nakamura, T. (2011). Determination of Cr(III) and Cr(VI) in water by wavelength-dispersive X-ray fluorescence spectrometry after preconcentration with an ion-exchange resin disk. *X-ray Spectrometry*, 40: 301-305.

[13] Hagiwara, K., Inui, T., Koike, Y., Aizawa, M., Nakamura, T. (2015). Speciation of inorganic arsenic in drinking water by wavelength-dispersive X-ray fluorescence spectrometry after in situ preconcentration with miniature solid-phase extraction disks. *Talanta*, 134: 739-744.

[14] Hagiwara, K., Koike, Y., Aizawa, M., Nakamura, T. (2016). Speciation of selenium in water by disk solid-phase extraction/X-ray fluorescence spectrometry. *Research Reports School of Science and Technology Meiji University*, 53: 33-38.

[15] Hagiwara, K., Koike, Y., Aizawa, M., Nakamura, T. (2016). On-site Determination of Trace Elemental Ions in Water Using Handheld X-ray Fluorescence Spectrometer. *Advances in X-ray chemical analysis, Japan*, 47: 249-255.

[16] Hagiwara, K., Koike, Y., Aizawa, M., Nakamura, T. (2015). On-site quantitation of arsenic in drinking water by disk solid-phase extraction/mobile X-ray fluorescence spectrometry. *Talanta*, 144: 788-792.

[17] Hagiwara, Kai, S., K., Koike, Y., Aizawa, M., Nakamura, T. (2016). On-site Determination of Heavy Metals in Drinking Water by Disk Solid-phase Extraction/Handheld X-ray Fluorescence Analysis. *Bunseki Kagaku*, 65: 489-495.

[18] Hagiwara, K., Koike, Y., Aizawa, M., Nakamura, T. Under article submission.

[19] Koike, Y., Hagiwara, K. (2015). Analysis of Radioactive Nuclides in Environmental Sample Using Solid Phase Extraction Disk. *Journal of the Society of Inorganic Materials, Japan*, 22: 408-413.

[20] Yasutaka, T., Shin, M., Onda, Y., Shinano, T., Hayashi, S., Tsukada, H., Aono, T., Iijima. K., Eguchi, S., Ohno, K., Yoshida, Y., Kamihigashi, H., Kitamura, K., Kubota, T., Nogawa, N., Yoshikawa, Y., Yamaguchi, H., Sueki, K., Tsuji, H., Miyazu, S., Okuda, Y., Kurihara, M., Tarjan, S., Matsunami, H., Uchida, S. (2017). Comparison of Concentration Methods for Low-level Radiocesium in Fresh Water. *Bunseki Kagaku*, 66: 299-307.

[21] Koike, Y., Sato, J., Nakamura, T. (2004). Gamma-ray spectrometry of ^{210}Pb in rainwater using preconcentration with iminodiacetate extraction disk. *Bunseki Kagaku*, 53: 1469-1473.

[22] Koike, Y., Sumiyama, H., Odagiri, Y., Inui, T., Iwahana, Y., Kurihara, Y., Nakamura, T. (2013). Gamma-ray Spectrometry of Short-lived Radionuclides in Rain Water Collected with Solid Phase Extraction Disk. *Bunseki Kagaku*, 62: 507-512.

[23] Koike, Y., Suzuki, R., Ochi, K., Hagiwara, K., Nakamura, T. (2017). Radioactivity Analysis Using Commercially Available Chemical Reagents as Calibration Sources, *Bunseki Kagaku*. 66: 263-270.

[24] Fukuda, D., Hagiwara, K., Suzuki, R., Kurihara, Y., Nakamura, T., Koike, Y. (2018). Preparation Method of Gamma-ray Disk Source Using Commercially Available Chemical Reagents. *Radioisotopes*, 67: 59-66.

In: Solid-Phase Extraction   ISBN: 978-1-53614-582-3
Editor: Ben Benson   © 2019 Nova Science Publishers, Inc.

Chapter 4

THE EVOLUTION OF SPE
SORBENTS IN A TIMELINE

Efstratios Agadellis, Artemis Lioupi
and Victoria Samanidou[*]
Laboratory of Analytical Chemistry, Department of Chemistry,
Aristotle University of Thessaloniki, Thessaloniki, Greece

ABSTRACT

Sample preparation always plays a key role in the chemical analysis
of a variety of matrices and is the most time-consuming and complicated
step of the entire analytical procedure. Solid-phase extraction (SPE) has
been one of the most widely applied innovative sample preparation
techniques, and since its introduction as an alternative to the liquid-liquid
extraction (LLE), it has been presenting numerous advantages. Its
benefits, compared to other techniques include simplicity and rapidity,
which are both significant for the modern analytical and green chemistry
demands, as well as selectivity, good repeatability and recoveries, low
limits of detection (LODs), use of low solvent amounts while no
specialized equipment is required.

---

[*] Corresponding Author Email: samanidu@chem.auth.gr.

The development of SPE has been quite rapid throughout the years and this book chapter examines the evolution of the used SPE sorbents since the introduction of the technique until nowadays.

1. INTRODUCTION

Modern analytical chemistry is based on the development of simple, environmentally friendly, sensitive, and selective procedures for the analysis of trace components including pre-concentration methods and additional determination by physical or physico-chemical methods [1]. One of the most significant steps in analytical processes is sample preparation due to its contribution in achieving low detection limits following legislation demands as well as eliminating matrix interference by clean up. Solid-phase extraction (SPE) is the most commonly used sample preparation technique for liquid as well as semi-liquid samples and has already replaced liquid–liquid extraction (LLE) in many protocols. In comparison to LLE, SPE is faster, it demands low solvent volumes and can handle low sample sizes of some micro- or milliliters. SPE has been widely applied in the purification and concentration of various compounds from complex matrices, such as environmental, food and biological samples [2, 3]. The first experimental application of SPE began 50 years ago, however this technique was not employed with analytical aims until the mid-1970s. Over the last few years, SPE experienced a series of changes such as the introduction of novel materials and sorbents formats, miniaturization and automation. Thus application range expanded to include more sample types. Modern extraction techniques have been developed such as solid-phase dynamic extraction (SPDE), micro-extraction by packed sorbent (MEPs), matrix solid phase dispersion (MSPD), stir-bar sorptive extraction (SBSE), solid-phase micro-extraction (SPME), automated headspace dynamic solid-phase extraction, and dispersive solid-phase extraction (d-SPE) [4].

Recently, the scientific community has put great effort into discovering novel materials that have higher sorption area and selective extraction properties. The extraction of e.g., highly polar compounds and macromolecules from aqueous samples still remains demanding. For this reason, many studies are being conducted to discover a proper substitution of non-specific (surface-modified silicas and porous polymers) sorbents for compound-specific and class-specific sorbents. Mixed-mode polymeric sorbents, surfactant-modified sorbents, molecular recognition sorbents (molecularly-imprinted sorbents, aptamers and immunosorbents) and nanostructured materials are some of the modern class-specific and compound-specific sorbents for SPE. Furthermore, efforts are being made to improve some properties of the materials such as mechanical, thermal and chemical stability as well as sorbents with advantageous particle size and morphology appropriate for higher mass-transfer rates [5].

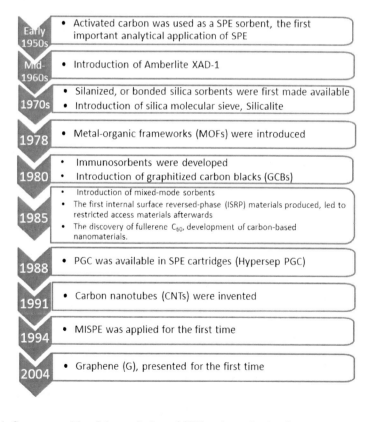

Figure 1. Summary table of the evolution of SPE sorbents in timeline.

Until now, novel SPE sorbents are being discovered and it appears to be a never-ending story. Nevertheless, nowadays efforts are targeted mostly at discovering an optimum sorbent for a specific utilization rather than a universal material suitable for every purpose. In this chapter, the evolution of sorbents in timeline applied in solid phase extraction has been summarized and discussed. The evolution of SPE sorbents in timeline is summarized in Figure 1.

2. SPE Sorbents and Applications

2.1. Activated Carbon

Historically, the first solid-phase extraction sorbent ever used was animal derived charcoal for removing pigments from chemical reaction mixtures. The charcoal was extracted with the compounds it had absorbed. However, the aim of this sample preparation technique is not to discard analytes of interest but to gather them and remove all interfering compounds which we do not wish to analyze [6]. Probably the first important analytical application of SPE was in the early 1950s and it is related to the use of metal pipes filled with activated carbon for the isolation of organic contaminants from various water sources [7]. Activated charcoal was firstly described as "any form of carbon capable of adsorption". The Roman and Chinese Empire first started using charcoal as a sorbent. The Romans realized that charcoal could purify water; a property we still utilize [8]. Activated carbon consists of carbon materials or char with increased surface area and its basic structural unit resembles the structure of pure graphite [9]. The main carbonaceous materials for the production of this sorbent are wood, coal, lignite and coconut shell. The large porosity, high surface area, well-developed internal pore structure consisting of micro-, meso- and macropores, as well as a wide spectrum of functional groups present on the surface of activated carbon render it a ''flexible'' material, which has various applications in different fields, but mainly in the environmental analysis [10]. In addition, it is important to note that activated carbon hardly binds inorganic constituents quantitatively at trace and ultratrace levels. As a result, several types of surface processes including oxidative and non-oxidative methods have long been employed to modify the interfacial region by increasing surface functional groups and hence, improving its potentially low adsorption capacity. Research teams have examined the non-oxidative surface modification, including loading, anchoring, immobilization, and grafting of organic molecules, which has gained considerable attention in recent years [11].

The simultaneous analysis of metals was examined by N. F. Al Qhadi and A. O. Al Suhaimi in 2017, who modified and prepared an activated carbon resin (AC-8-HQ) for the determination of five metal elements in groundwater samples (r Mn(II), Cd(II), Ni(II), Pb(II) and Zn(II)). The solid-phase preparation was based on the immobilization of the generated p-nitroaniline aryldiazonium salt on the surface of activated carbon, with the presence of hypophosphorous acid. The newly prepared resin worked within an appropriate pH range of 5.5-6.5 [12].

Tire-based activated carbon materials was employed by K. M. Dimpe et al. in 2018, for the simultaneous determination of several metal ions in domestic and industrial wastewater. The tire-based sorbent was prepared by rinsing and then drying for 24 h at 100°C, while the method was optimized by two-level full factorial design (FFD), where the simultaneous variation of the analytical parameters allowed a cost-effective method development. The sorbent was characterized by scanning electron microscope (SRM) presented irregular morphology, with various shapes of the particles. The optimization presented satisfactory limits of detection (between 0.66 - 2.12 µg L^{-1}), while the recoveries ranged from 92 to 99% [13].

2.2. Polymeric Sorbents

In the mid-1960s, Rohm and Haas Company introduced a cross-linked polystyrene resin, Amberlite XAD-1, as a replacement for active charcoal.

The resin was in form of beads (20 – 50 mesh), each of which was a conglomerate of a large number of microbeads. The series of Amberlite polymeric resins started to attract the attention of other analysts in the early 1970s. Apart from XAD-1, other styrene–divinylbenzene Amberlites (XAD-2, XAD-4) and ethylene–dimethacrylate resins (XAD-7 and XAD-8) were also introduced [14]. These resins have frequently been applied in Solid Phase Extraction, on account of the good thermal stability and high sorption capacity they offer toward ions with better flexibility in working conditions. However, some of the disadvantages of these materials are low selectivity and regenerability [15]. Amberlite XAD resins are very porous spherical polymers based on highly crosslinked, macroreticular polystyrene, aliphatic, or phenol-formaldehyde condensate polymers. On the basis of polymeric matrix, they are separated into two main groups: i) polystyrene-divinyl benzene-based resins including XAD-1, XAD-2, XAD-4, XAD-16, XAD-1180, XAD-2000 and XAD-2010 and ii) polyacrylic acid ester-based resins including XAD-7, XAD-8 and XAD-11 [16].

Highly crosslinked polymers achieve strong π-π interactions between analytes and sorbents. Furthermore, their open structure enhances their specific surface area. Lichrolut EN, Envi-Chrom P, Styrosorb, Isolute ENV, and HYSphere-1 are some highly crosslinked polymers, which achieve higher recoveries than the ordinary sorbents in the trace enrichment of polar analytes from aqueous samples [17].

Novel functionalized polymeric sorbents like Oasis (Waters, MA, USA), Strata (Phenomenex, Torrance, USA), Bond Elute PPL (Varian, CA, USA), LiChrolut EN (Merck, Darmstadt, Germany), etc. have been developed combining hydrophilic and lipophilic properties. Among these polymeric materials the most commonly used in solid phase extraction is the co-polymer of polystyrene divinylbenzene. In 1996, Oasis HLB was introduced as a product of polymerization of lipophilic DVB and hydrophilic N-vinylpyrrolidone, making a hydrophilic – lipophilic balance. This sorbent has been applied for SPE of polar to apolar organic compounds. Strata-X is a surface modified polymer of styrene and divinylbenzene. It should also be noted that this sorbent retains analytes during SPE by hydrophobic, hydrophilic and $\pi - \pi$ interaction. These characteristics render Strata X a universal SPE material for basic, neutral and acidic analytes. These polymeric sorbents require a small quantity of sorbent due to their high capacity and they are used as an alternative to RP silica. Moreover, macroporous and wettable materials present a straightforward adsorption mechanism and do not suffer from such problems like metallic impurities, pH stability or active silanol groups, which arise when using RP silica as SPE sorbent [18, 19]. This dual character permits the use of these materials in a way that conditioning step to prepare sorbent for sample contact is not necessary.

A novel β-cyclodextrin microporous polymer (MP-CDP) was tested in 2018 by Y. Li et al. as a solid-phase extraction sorbent for the simultaneous detection of trace amounts of bisphenols (BPF, BPA, BPAF) in water samples and orange juices [20]. MP-CDP was prepared with sequential additions of materials and degassing with nitrogen. The acquired material was dried at room temperature for 3 days. Decafluorobiphenyl was used as the cross-linker. Following its manufacturing, the sorbent was characterized after a series of processes, such as Fourier transform infrared spectroscopy, solid-phase ^{13}C nuclear magnetic resonance (NMR), thermogravimetric analysis, scanning electron microscopy (SEM), and nitrogen adsorption and desorption analysis. Its extraction efficiency, as well as its adsorption ability were proved as more than satisfactory for the determination of bisphenols with HPLC-UV, with limits of detection of 0.15 ng mL^{-1} and 0.5 ng mL^{-1} for water samples and orange juice respectively. Recoveries were 95.7 – 106.3% for BPF, 92.9 – 107.0% for BPA, and 96.0 – 103.5% for BPAF [21].

2.3. Silica Based Sorbents

The SPE started being widely used in the late 1970s, when the application of silanized, or bonded silica sorbents, was first made commercially available [13]. Although silica is commonly represented by a simple molecular formula (SiO_2), in reality, it is an inorganic polymeric material. In silica, tetrahedral structures are formed by silicon and oxygen atoms with the silicon atom at the center, and the four oxygen atoms on the four corners of each tetrahedron. The most commonly used in chromatographic separations and sample preparations is amorphous silica, in which each of these tetrahedra may be chemically attached to up to four neighboring tetrahedra through silicon-oxygen (siloxane) bonds in a random way [22]. In 1978, the first article using SPE on a bonded phase silica was published, which described the use of a Sep Pak™' C_{18} "cell" for the cleanup of histamines from wines. The new prepackaged, disposable cartridges/columns containing bonded silica sorbents, which were introduced in 1977, demonstrated that silica rapidly equilibrates with sorbent, has a high surface area and is relatively stable to changes in temperature and pH within a range of 2 to 7. However, the properties of these types of sorbents do not seem to change when they are exposed to liquids with a pH outside this range for short periods [23, 24]. The first SPE silica-based materials were modified with C_{18}, C_8, CH, CN, phenyl, or NH2 groups. The main disadvantages of these materials are instability at extreme pH, low recovery in the extraction of polar analytes, and the presence of residual silanol groups [25]. According to the chemical groups bonded to the silica, the phases are classified as non-polar, polar or ion-exchangers. Octadecyl surface phases (C_{18}) are used for the reverse-phase extraction of nonpolar compounds in aqueous solution. The shorter octyl phases (C_8), are used to extract medium polarity compounds, while silica gel and alumina oxides are used for extracting polar compounds. The first pure silica molecular sieve, Silicalite, was synthesized in the 1970s. Silicalite is a polymorph of SiO_2 which exhibits a high degree of organophilichydrophobic character. It is capable of separating organic molecules out of water-bearing streams. The pores of Silicalite are microporous, 6 Å in diameter, which give the molecular sieve size exclusion properties. Molecules small enough to enter the channels are retained through hydrophobic interactions [26].

Silica-based sorbents have been used for the extraction of compounds from water samples, as well as more complex ones, such as food samples. [27, 28] In 2018, N. Casado et al. studied and characterized several mesostructured silicas with wormhole-like pore arrangement as sorbents for the simultaneous analysis of a number of veterinary drugs residues. The characterization of those prepared materials took place with transmission electron microscopy (TEM), scanning electron microscopy (SEM), nitrogen adsorption-desorption isotherms, elemental analysis and thermogravimetric analysis (TGA). Initially, the less effective sorbents for the target compounds were SBA-15 and MCM-41, while ethane-PMO showed low recoveries for analytes such as β-agonists, β-blockers and NSAIDs. The most satisfactory results were obtained with HMS-C$_{18}$, with the best adsorption and recoveries. HMS-C18 is characterized by irregular structure uniform, with mesopores with uniform pore sizes, as well as wormhole-like pores, which may cause a bottleneck effect. The sorbent meets some of the Green Chemistry principles, as it may be prepared with lower costs and it is environmentally friendly. Its comparison with the rest mesoporous silica sorbents showed that it presents remarkable advantages over all the rest examined sorbents, like more regular flow and better extraction efficiency due to spherical morphology of its particle sizes, which results in more effective SPE packaging. In addition, it presents better adsorption and extraction efficiency in a greater variety of compounds, with versatile polarity and chemical structure [27].

As previously referred, there have been applications of silica-based materials in water samples too. Recently, E. Pellicer-Castell et al. [28] prepared and evaluated mesoporous silica materials doped with Ti and Fe metals respectively, as well as immobilized cyclodextrin silica-based supports (β- or γ-CDs). The surface area of the sorbents was studied with the nitrogen adsorption-desorption isotherms and presented bimodal framework with two adsorption steps. Both pure and doped with Ti and Fe mesoporous models presented both mesopores of 2.8 nm and larger pores (33-39 nm). The study of CD-silica composite materials showed lower surface areas and smaller pore sizes, caused by the addition of CDs. The optimum results were obtained with the mesoporous silica material doped with Ti (Ti25-UVM-7) with recoveries ranging from 84% to 104.5% and limits of quantification between 0.5 - 4.4 µg L^{-1}. The solid-phase extraction procedure was coupled to mass spectrometry detection and a cost-effective and environmentally friendly method was developed.

2.4. Metal-Organic Frameworks

The novelty of metal-organic frameworks (MOFs) was introduced in 1978 with a small number of structures and application in its early years. However, the interest of the scientific community has increased exponentially over the last decade, with plenty of publications introducing newly-developed and structured materials used in chemical analysis [29, 30]. MOFs are composed of metal ions connected with organic ligands and have been used for the separation and determination of several compounds from environmental and biological samples. They are characterized by abundant surface area and porosity, as well as versatile structure, topology and porosity. Their structure can be modified regarding the target compound and can affect the SPE efficiency. When it comes to selectivity, this can be enriched with interactions (π-π, hydrophobic) between the target compounds and the sorbent's framework, as well as their porosity and shape are important for the efficient adsorption of the analyte on the material's surface.

In 2011, Gu et al. [31] introduced a method for the determination of peptides from complex biological samples (urine or human plasma), with adsorbents such as MIL-53, MIL-100, and MIL-101. With the addition of the adsorbents in the samples and following incubation, the samples were analyzed after removal of the supernatant and by matrix-assisted laser desorption/ionization time-of-flight-mass spectrometry (MALDI-TOF-MS).

Furthermore, in 2018, S. Zhang et al. [32] prepared magnetic zinc adeninate metal-organic frameworks, that were utilized for the extraction of 6 benzodiazepines from urine and wastewater. For the preparation of the adsorbents, Fe_3O_4 nanoparticles were put onto the external surface of cobalt adeninate metal–organic frameworks by using amino-silane as a linkage (Fe_3O_4-NH_2/bio-MOF-1). The adsorption of the analytes on the

MOFs surface was achieved by hydrogen binding, π-π interactions and electrostatic attraction. The achieved limits of detection were between $0.71 - 2.49$ ng L^{-1}.

Table 1 summarizes recent applications of MOFs in solid-phase extraction (SPE).

2.5. Mixed-Mode Sorbents

In 1985, mixed-mode (reversed phase/ion exchange) sorbents were introduced [7]. There are several types of mixed-mode sorbents: cationic or anionic and strong or weak ion-exchange, depending on the ionic group attached to the resin. These groups are created to selectively extract analytes with special chemical properties (i.e., strong/weak acidic or basic). It is important to note also, that these materials are used to selectively extract analytes (charged or uncharged) from complex biological, food or wastewater matrices. Over the past few years, mixed mode technology has attracted the attention of the scientific community. Thus far, several new materials have been developed due to the interest for cleaner extracts from SPE. This demand was especially driven for preventing the ion suppression/enhancement when the SPE extracts are injected into liquid chromatography-mass spectrometry (LC-MS) systems [40].

Mixed-mode sorbents have been utilized in various analyses, but they have recently seen several applications in the extraction and determination of non-steroidal acidic inflammatory drugs (NSAIDs), some of the most well-known and consumed by both humans and animals. For instance, C. Huang et al. introduced in 2018 [41] a new mixed-mode polymeric sorbent, the PDVB support, which was synthesized by copolymerization of divinylbenzene and 2-(diethylamino)ethyl methacrylate via Pickering emulsion polymerization. In the sequel, the quaternized hyperbranched macromolecules (QHMs) were grafted on the PDVB's surface with sequential reactions of resorcinol diglycidyl ether with methylamine (N, N-dimethylethanolamine for terminal epoxides). This is a high-capacity mixed-mode anion-exchange (MAX) sorbent and in this work, it was used for the analysis and determination of non-steroidal acidic anti-inflammatory drugs (NSAIDs). The variation of specific surface area (SBET), pore volume and ion exchange capacity (IEC) with generation number reveal that the QHMs have been grown successfully within the large meso-channels of the porous aminated PDVB. An efficient approach based on the mixed-mode SPE coupled with HPLC-UV was developed for highly selective extraction and cleanup of nine NSAIDs (tolmetin, ketoprofen, naproxen, flurbiprofen, diclofenac, indomethacin, ibuprofen, mefenamic acid, tolfenamic acid) in human urine. The hyperbranched MAX for sample analysis of NSAIDs, with recoveries from 81.9 to 104.0% and detection limits of 0.004 – 0.009 µg mL^{-1} was proved as an appropriate sorbent for the determination of complex compounds such as NSAIDs.

In 2017, Y. Li et al. [42] also worked on four NSAIDs (ketoprofen, naproxen, diclofenac, and ibuprofen) and their extraction with a novel reversed-phase/anion-exchange mixed-mode silica sorbent in water samples (tap water, river water, and wastewater). Initially, a mesoporous silica Santa Barbara Amorphous-15 (SBA-15) was modified with a silane with three amines (3-[2-(2- aminoethylamino)ethylamino]propyl-trimethoxysilane) and in the sequel, it reacted with phenyl glycidyl, in order to acquire both reversed-phase and anion-exchange modes. As a result of its characterization, the sorbent presented satisfactory surface area, pore volume, particle sizes, all of which made them appropriate for the NSAIDs determination method development. The mixed-mode SPE-HPLC-UV method reached at low limits of detection (0.006 – 0.070 g L^{-1} for tap water, and 0.014 – 0.16 g L^{-1} for river water and wastewater) and recoveries of 80.6% - 110.9%.

Table 1. Recent applications of MOFs in solid-phase extraction (SPE)

Analyte	Sorbent	Matrix	Detection system	LOD	PF	Ref.
As	Fe_3O_4@ZIF-8	Urine	HG-AFS	0.003		[33]
Benzodiazepines	Fe_3O_4-NH_2/bio-MOF-1	Urine	LC-MS	0.00071 – 0.00249	37/102	[32]
Chrophenols	Fe_3O_4@MOF-5-C	Mushroom	HPLC-UV	$0.25 – 0.30$ ng g^{-1}		[34]
NAP and D-NAP	MIL-101	Urine	HPLC-Flu	0.011 - 0.034	290 - 295	[35]
OPPs[a]	Fe_3O_4@MIL-100 (Fe)	Hair and urine	GC-Fid	0.00021 - 2.28		[36]
Pthalate esters	Fe_3O_4@MIL-101	Plasma	GC-MS	0.08 - 0.15		[37]
Sex hormones	MIL-53-C	Water/Urine	LC-MS	0.1 - 0.3		[38]
THC	Chitosan@MIL-101	Blood	HPLC-UV	0.04	970	[39]

[a] Organophosphorous pesticides

2.6. Graphitized Carbon Blacks (GCB's) and Porous Graphitic Carbon (PGC)

The modern forms of carbon used for SPE are graphitized carbon blacks (GCB's) and porous graphitic carbon (PGC). In particular, the Porous Graphite Carbon technology (PGC) was developed by Knox in 1982, using the reverse template process [43]. In 1988, PGC was available in SPE cartridges (Hypersep PGC) and was similar to the LC-grade Hypercarb. It is believed that PGC contains residual acidic and basic functional groups at its surface which are related to its ability to bind polar and non-polar compounds [44].

The use of carbonaceous sorbents for SPE began in the 1980s with the introduction of graphitized carbon blacks (GCBs) obtained by heating carbon blacks at high temperature (2700 – 3000°C). The first available GCBs were non porous with a low specific surface area around 100 m²/g (Carbopack B or ENVI-Carb SPE from Supelco, Carbograph 1 from Altech). Carbograph 4 was recently introduced with a surface area of 210 m²/g. These sorbents were used for the extraction of non-polar analytes such as organochlorinated insecticides and gained attention when it was proved that they isolate polar molecules with high solubility in water (> 1 g/l). GCB was proved to contain several functional groups at the surface following the oxygen chemisorption and it was applied in the extraction of organochlorinated insecticides and other non-polar analytes [45].

2.7. Carbon Based Nanomaterials

The development of carbon-based nanomaterials has been one of the most significant trends in solid phase extraction, after the discovery of fullerene C_{60} in 1985. Lately, a great amount of these materials have been researched as sorbents in sample pretreatment. These materials include carbon nanotubes, carbon nanocones-disks and nanohorns, carbon nanofibers, fullerenes, graphene, as well as their functionalized forms [46]. Among them, the most extensively applied materials in analytical chemistry are carbon nanotubes (CNTs), which were invented by Iijima in 1991, and graphene (G), which was presented for the first time by Novoselov and Geim in 2004. Graphene is an allotropic form of carbon with a thin (one atom thick) planar formation of sp2-bonded carbon atoms. On the contrary, CNTs are hollow graphitic materials composed of one or multiple layers of graphene sheets (single-walled carbon nanotubes, SWCNT, and multi-walled carbon nanotubes, MWCNT, respectively). The two materials have much the same carbon-based formation and consequently they have similar optical, mechanical, electric and magnetic characteristics, including an extremely high surface/mass ratio. It is of interest that CNTs and graphene have been utilized for various applications in analytical chemistry. In particular, they have been used in biosensors and microelectrodes, as stationary phases in chromatography, as pseudostationary phases in capillary electrophoresis, etc. Nevertheless, most of these materials are employed as sorbents in Solid Phase Extraction because of the extremely high surface area, the efficiency to establish analyte-sorbent bonds and the ability of surface qualification by including selective functional groups [47].

Their modification versatility in order to be compatible with different types of solvents is the main reason for CNTs to have been implemented in many applications in recent years. For instance, for the analysis of heavy metals, Baghban et al. introduced in 2012 an SPE coupled with flame atomic absorption spectrometry (FAAS) method for the extraction of lead (II) and cadmium (II), utilizing alizarin red S modified TiO_2 nanoparticles. The method was based on the preconcentration of the two metals on the surface of a microcolumn packed with the sorbent and their extraction with hydrochloric acid. Following optimization of all analytical parameters, the method reached limits of detection of 0.11 and 0.30 ng mL^{-1} for Cd(II) and Pb(II) respectively, high accuracy and precision [48].

Table 2 presents main Contemporary applications of CNTs in solid-phase extraction (SPE).

Table 2. Contemporary applications of CNTs in solid-phase extraction (SPE)

Analyte	Sorbent	Matrix	Detection system	LOD (μg L^{-1})	PF	Ref.
Amoxicillin	Magnetic MWCNTs	Amoxicillin samples/Urine	Spectrophotometric	3.0 ng mL^{-1}		[49]
Antidepressants	CNTs incorporated polymer monolith	Urine	HPLC-UV	9 - 15		[50]
Cd	MWCNTs	Urine	ETAAS	0.01	3.4	[51]
Cd (II)	MWCNTs	Food and water	FAAS	0.70	100	[52]
Cu (II)				1.2		[52]
Ni (II)				0.80		[52]
Pb (II)				2.6		[52]
Zn (II)				2.2		[52]
Fluoxetine	Fe$_3$O$_4$@MWCNTs	Urine	UV-vis	60	48	[53]
Neonicotinoids, Sulfonylureas	Magnetic CNTs – C$_{18}$-modified nano SiO$_2$	Water/Fruit Juices	HPLC-UV	0.07 to 0.60 μg mL^{-1}		[54]
Sc (II)	oxidized carbon nanotubes	natural water	ICP-OES	0.05	250	[55]
Zn	MWCNTs	Human hair, Water	FAAS	0.00007	250	[56]

2.8. Restricted-Access Materials

The RAMs used today originated from the first sorbents produced in 1985 by Hagestam and Pinkerton, called internal surface reversed-phase (ISRP) materials. Restricted access materials are separated into five basic types depending on the nature of the barrier and the surface structure of the sorbent: mixed-functional phases and dual-zone materials, shielded hydrophobic phases, semi-permeable surfaces, internal surface reversed-phase packings and polymeric materials [57, 58]. The name restricted-access material (RAM) was first used in 1991 by Desilets et al. These sorbents have special properties, based on a molecular-weight cut-off and they are able to fractionate a biological sample into the protein matrix and the analyte fraction. Concurrently to this size-exclusion process, low molecular-weight compounds are extracted and enriched via partition, into interior of the phase. The surface of the particles is in exposed to compounds such us nucleic acids, humic material and proteins and their structure prevents adsorption of these macromolecules. The elimination of macromolecules can be achieved by a physical barrier as a result of the pore diameter or by a chemical diffusion incited by a protein (or polymer) network at the exterior surface of the particle. Consequently, RAM sorbents that have a biocompatible surface and adsorption centers for the extraction of small analytes, are viewed as bifunctionally modified particles [59, 60]. Furthermore, it is speculated that numerous other sorbents (e.g., polymers, alumina, modified and unmodified silica, and natural sorbents like rice straw, among others) can be converted to the RAMs, given the fact that simple modifications are required on their outer surfaces [61].

When it comes to more contemporary applications, in 2015, J. He et al. coupled restricted access materials (RAMs) with molecularly imprinted polymers (MIPs), using malathion as a template and glycidilmethacrylate (GMA) as a co-monomer. The introduced sorbent was characterized by scanning electron microscopy. Its comparison with conventional non-fuctionilized MIPs, as well as restricted-access-non-molecularly imprinted polymers (RAM-NIPs) and their selectivity and adsorption capacity for the target analytes was higher. The RAM-MIP was used for the extraction of organophosphorous pesticides from honey samples. That new SPE process was combined with gas chromatography and achieved simultaneous detection of 6 pesticides (malathion, ethoprophos, phorate, terbufos, dimethoate, and fenamiphos) with low detection limits ($0.0005 – 0.0019$ μg mL^{-1}). In addition, the SPE-GC method provided simplicity, good performance of sample treatment and speed [62].

In 2018, V.M.P. Barbosa et al. prepared and used a restricted-access carbon nanotube (RACNTs) column for the determination of lead (II) in serum samples, without any previous treatment required. The material emerged after the coupling of commercial carbon nanotubes with bovine serum albumin (BSA) layers. The newly prepared material was characterized by transmission electron microscopy, scanning transmission electron microscopy and energy dispersive spectroscopy, which showed its Pb^{2+} adsorption capacity and sites. The method reached a limit of detection of 2.1 mg L^{-1} and recoveries from 89.4% to 107.3% [63].

RACNTs are promising materials in terms of adsorption and extraction of compounds like metals and macromolecules from complex matrices, like biological fluids and food samples [62, 63].

2.9. Molecularly-Imprinted Sorbents/Ion-Imprinted Sorbents

In 1940, Pauling introduced molecular impression as a possible explanation for the generation of antibodies in living systems triggered by antigens [64]. In recent years, MIPs have been widely used as SPE sorbents, namely, in molecularly imprinted solid-phase extraction (MISPE), due to their high selectivity. Sellergren first applied MISPE in 1994, for the extraction of pentamidine from urine. More recently, it has been used in the selective extraction or cleanup of target analytes from various complex (food, biological and environmental) samples [64]. Furthermore, the structure of MIPs allows selective interactions with the desired analyte, reducing matrix effects [65]. Rebinding interactions, the size and the shape of the cavity are some factors that determine the selectivity of the molecularly imprinted polymer. These sorbents are stable during chemical and physical procedures, including organic solvents, heating, bases and acids. They can be regenerated and repeatedly used without memory losses of molecular recognition sites and they can be stored for several years in a dry state and at room temperature. Target analytes are inexpensive, stable and easy to prepare which make them suitable for various applications [66].

As for some recent developments of MIPs, J. Yang et al. introduced in 2015 a new method for the simultaneous determination of eight bisphenols (A and its structural analogues F, B, E, AF, AP, Z and S) with molecular imprinted polymer microsphere particles (MIPMS) in human urine samples. The particles were prepared with the process of Pickering emulsion polymerization, which provides better control of particle size and higher yields of polymer. The produced polymers met all the requirements for microspherical shape, surface area, total pore volume, selectivity and adsorption capacity for all 8 bisphenols. The analysis with HPLC coupled with diode array detection (HPLC-DAD) achieved detection limits between 1.1 and 2.2 ng mL^{-1} and the recoveries between 81.3% and 106.7% [67].

Moreover, in a 2016 work, a MISPE was applied for the isolation and determination of ampicillin sodium from milk and blood samples. This MIP was prepared through a surface imprinting process, where a MIP shell was anchored at a silica gel surface prior to HPLC analysis. The developed method reached low LOD (0.15 μg mL^{-1}), after separation and determination with HPLC [68].

Table 3 summarizes recent advances in MIPs' applications as SPE sorbents.

The ion-imprinted sorbents were initially introduced in 1976 by Nishide et al. for the determination of metal ions. The production of IIPs is based on the reaction of a metal ion and a ligand, forming together a complex [85]. In recent years, there have been successful attempts for merging MIPs/IIPs with nanoparticles such as magnetic nanoparticles and CNTs for the improvement of the extraction ability and the selectivity of the materials. The results showed better adsorption capacity of the newly-produced sorbent. Following the material's polymerization and by using a proper solvent such as a mineral acid, the new imprinted polymer is collected [29].

According to works that have been published, the IIPs have shown exceptional extraction efficiency and selectivity in the determination of metal ions from matrices like milk, water, biological, and other environmental samples [87-93].

Recent advances in IIPs' applications as SPE sorbents are presented in Table 4.

Table 3. Recent advances in MIPs' applications as SPE sorbents

MIPs					
Analyte	**Sorbent**	**Matrix**	**Detection system**	**LOD (μg L^{-1})**	**Ref.**
Amphenicol antibiotics	MDMIP	Blood	HPLC-UV	0.08 - 0.16 μg kg^{-1}	[69]
Ampicillin sodium	SMIP	Milk, Blood	HPLC-UV	150	[68]
Baicalein	SBA15@MIP	Plasma	HPLC-UV	3.5	[70]
Biochanin A	SiO$_2$@MIP	Urine	HPLC-DAD	0.04 - 0.06	[72]
Bisphenols	MIPMS	Urine	HPLC-DAD	1.2 - 2.2	[67]
Carvediol	MMIP	Serum	HPLC-UV	0.13	[71]
Dicoumarol	MIP	Plant sample	HPLC-DAD	200	[73]
Gatifloxacin	MCNTs@MIP	Serum	HPLC-UV	6.0	[74]
Haloperidol	MIP	Plasma, Urine	HPLC-DAD	0.3 - 0.35	[75]
Insulin	MIP	Plasma, Urine	HPLC-DAD	0.2 - 0.7	[76]
Ketoprofen	MIP	Wastewater	HPLC-UV	0.09 - 0.23	[77]
Lamotrigine	MMIP	Plasma, Urine	HPLC-UV	0.5/0.7	[78]
NSAIDs	MIP	Urine	HPLC-DAD	300 - 400	[79]
Paracetamol	MMIP	Plasma	HPLC-UV	0.17	[71]
Parathion	MIP	Apple	CLEIA	0.1 μg kg^{-1}	[80]
Parathion		Rice		0.12 μg kg^{-1}	[80]
Parathion		Orange		0.21 μg kg^{-1}	[80]
Parathion		Cabbage		0.05 μg kg^{-1}	[80]
Progesterone hormones	MIP	Blood and urine	GC-FID	0.625	[81]
Rhodamine B	Fe$_3$O$_4$@SiO$_2$-NH$_2$	Food samples	HPLC-UV	3.4	[82]
Sitagliptin	MIP	Plasma, Urine	HPLC-DAD	0.03	[83]
Sterols	MIP	Serum	GC-MS	1.1 - 3.6	[84]

Table 4. Recent advances in IIPs' applications as SPE sorbents

IIPs						
Analyte	Sorbent	Matrix	Detection system	LOD ($\mu g\ L^{-1}$)	PF	Reference
Cd	Cd-IIP-MMS	Urine	GFAAS	0.0061	50	[87]
Co	$Fe_3O_4@TiO_2@SiO_2$-IIP	Urine	FAAS	0.15		[88]
Cu (II)	Cu-IIP	Human hair	ETAAS	0.0087	100	[89]
Hg	Hg-IIP	Human hair	AFS	0.015		[90]
Pb	IIHC	Urine	FAAS	0.75	128	[91]
Pb	MIIP	Human hair	FAAS	0.01	150	[92]
Zn	IIP	Cereal	GFAAS	0.48	118	[93]
Zn	IIP	Water	GFAAS	1.02	28	[93]

2.10. Immunosorbents

Immunosorbents (IS) were developed to take advantage of the specificity of antibodies for the isolation of target analytes from complex mixtures. In the early 1980s, biomedical applications developed followed by specific procedures for food contaminants and subsequently environmental applications by the mid-1990s [7]. The immunoaffinity extraction is based on the interaction between antigen–antibody. Antibodies produced against a particular compound can be immobilized by covalent bonding to the surface of an appropriate support or, alternatively, they can be encapsulated into the pores of a solid matrix. These biomaterials will selectively hold the analyte–antigen present in, for example, a surface water sample, thus achieving extraction, preconcentration, and cleanup in the same step. Nevertheless, due to the inevitable cross-reactivity of antibodies, particularly when they are generated against a little molecule with few determinant groups (as is the case for nearly all pesticides), other structurally related compounds may also be retained by the IS. This cross-reactivity has contributed to developing new class-selective sorbents that can be applied for the simultaneous SPE of various members of the same chemical family. It has been remarked that sample preparation methods utilizing IS also led to increased sensitivity of analysis for complex matrices because, as the obtained extracts are theoretically free of most matrix interferences, it is likely to employ highly sensitive detection conditions [94].

M. Bonichon et al. developed in 2018 an easy and quick method for the determination of human butyrylcholinesterase (HuBuChE) [95], which is an indicating biomarker of the exposure to organophosphorous to warfare agents, in low quantities of human plasma. The solid-phase extraction was coupled with micro liquid chromatography-tandem mass spectrometry. In trial, three anti-HuBuChE antibodies were grafted on CNBr-sepharose and epoxy-polymethacrylate supports and packed in precolumns. Sepharose-based supports, with grafting yields up to 98% presented the best results, while the B2 18-5 monoclonal antibody grafted on sepharose led to the optimum immunosorbent, with high HuBuChE recoveries, nearly 100%. During the method development, on-line digestion set-up and immunoextraction of HuBuChE was achieved in 14 min, while digestion was performed in 20 min, allowing detection of the target nonapeptide in less than 1 h. The global recovery of the nonapeptide was higher than 42% using the best immunosorbent (B2 18-5$_{SP}$). The obtained limit of quantification of the target nonapeptide in human plasma sample was 2 fmol, which is below the average amount of HuBuChE tetramer in 50 L of plasma (590 fmol). Remarkable interferences could be avoided with that extraction method, with the removal of albumin from plasma during extraction and washing.

CONCLUSION

Solid-phase extraction has been, is and will be a sample preparation technique of utmost importance. The variety and the continuous evolution of the materials used in timeline, which is presented in the above-mentioned text is going to still concern for a long time the scientific community and the field of Analytical Chemistry with plenty of applications that will achieve cost-effectiveness, speed, low limits of detection, as well as high precision and accuracy.

activated carbon for its water treatment applications, *Chem. Engin. Journ.*, *219*, 499–511.

11]   Zhenhua, Li, Xijun, Changa, Xiaojun, Zoua, Xiangbing, Zhua, Rong, Niea, Zheng, Hua. & Ruijun, Li. (2009). Chemically-modified activated carbon with ethylenediamine for selective solid-phase extraction and preconcentration of metal ions, *Analytica Chimica Acta*, *632*, 272–277.

[12] Nawader, F. AlQadhi. & Awadh, O. AlSuhaimi. (2017). Chemically functionalized activated carbon with 8-hydroxyquinoline using aryldiazonium salts/diazotization route: Green chemistry synthesis for oxins-carbon chelators, *Arabian Journal of Chromatography* (Article in Press).

[13] Mogolodi Dimpe, K., Ngila, J. C. & Philiswa, N. Nomngongo. (2018). Preparation and application of a tyre-based activated carbon solid phase extraction of heavy metals in wastewater samples, *Physics and Chemistry of the Earth, Parts A/B/C*, *105*, 161-169.

[14] Liska, I. (2000). Fifty years of solid-phase extraction in water analysis-historical development and overview, *J. Chrom. A*, *885*, 3–16.

[15] Asif, Ali Bhatti, Shahabuddin, Memon, Najma, Memon, Ashfaque, Ali Bhatti. & Imam, Bakhsh Solangi. (2017). Evaluation of chromium(VI) sorption efficiency of modified Amberlite XAD-4 resin, *Arabian Journal of Chemistry*, *10*, S1111–S1118.

[16] Akil, Ahmad, Jamal, Akhter Siddique, Mohammad, Asaduddin Laskar, Rajeev, Kumar, Siti, Hamidah Mohd-Setapar, Asma, Khatoon. & Rayees, Ahmad Shiekh. (2015). New generation Amberlite XAD resin for the removal of metal ions: A review *Journal of Environmental Sciences*, *31*, 104-123.

[17] Nuria, Masque, Marina, Galia, Rosa, M. Marce. & Francesc, Borru (1999). Functionalized Polymeric Sorbents for Solid-Pha Extraction of Polar Pollutants, *J. High Resol. Chromatogr.*, *22*, (547–552.

[18] Muhammad, Nasimullah Qureshi, Guenther, Stecher, Chris Huck. & Guenther, K. Bonn. (2011). Preparation of polymer }

REFERENCES

[1]   Ali Rehber, Turker. (2007). New Sorbents for Solid-Phase E for Metal Enrichment, *Clean, 35* (6), 548 – 557.

[2]   Fontanals, N., Marce, R. M. & Borrull, F. (2007). New ma sorptive extraction techniques for polar compounds, *J Chrom. A, 1152,* 14–31.

[3]   Huck, C. W. & Bonn, G. K. (2000). Recent developr polymer-based sorbents for solid-phase extraction, *Journ. oj A, 885,* 51–72.

[4]   Auréa, Andrade-Eiroa, Moisés, Canle, Valérie, Leroy-Cance Víctor, Cerdà. (2016). Solid-phase extraction of organic com A critical review (Part I), *Trends in Analytical Chemistry, 8* 654.

[5]   Martina, Hákováa, Hedvika, Raabováa, Lucie, Chocho Havlíkováa, Petr, Chocholouša, Jiří, Chvojkab. & Dalibor, Šai (2018). Testing of nylon 6 nanofibers with different surface d as sorbents for solid phase extraction and their selectivity com with commercial sorbent, *Talanta, 181,* 326–332.

[6]   Thurman, E. M. & Mills, M. S. (1998). *Solid-Phase Extr Principles and Practice,* Wiley & Sons: New York, 147.

[7]   Colin, F. Poole. (2002). Chapter 12: Principles and practice of phase extraction, *Comprehensive Anal. Chem., 37,* 341-387.

[8]   Nikolas, Hagemann, Kurt, Spokas, Hans-Peter, Schmidt, Ralf, Marc, Anton Böhler. & Thomas, D. Bucheli. (2018). Acti Carbon, Biochar and Charcoal: Linkages and Synergies a Pyrogenic Carbon's ABCs, *Water, 10*(2), 182.

[9]   Afrida, Kurnia Putri, Wang-Hsien, Ding. & Han-Wen, Kuo. (2 Characteristic of New Solid-Phase Extraction Sorbent: Acti Carbon Prepared from Rice Husks under Base Treated Conditic *Pure App. Chem. Res., 1* (1), 1-10.

[10] Amit, Bhatnagar, William, Hogland, Marcia, Marques. & N Sillanpää. (2013). An overview of the modification method:

sorbents for solid phase extraction of polyphenolic compounds, *Cent. Eur. J. Chem.*, *9*(2), 206-212.

[19] Nader, Rifai, Rita Horvath, A., Carl T. Wittwer & Andy Hoofnagle. (2018). *Principles and Applications of Clinical Mass Spectrometry: Small Molecules, Peptides, and Pathogens, (III), 80.*

[20] Yarong, Lia, Pengpeng, Lua, Jincheng, Chenga, Xiaoliang, Zhua, Wang, Guoa, Lijun, Liua, Qiang, Wanga, Chiyang, Hea. & Shaorong, Liu. (2018). Novel microporous β-cyclodextrin polymer as sorbent for solid-phase extraction of bisphenols in water samples and orange juice, *Talanta*, *187*, 207-215.

[21] Yarong, Lia, Pengpeng, Lua, Jincheng, Chenga, Xiaoliang, Zhua, Wang, Guoa, Lijun, Liua, Qiang, Wanga, Chiyang, Hea & Shaorong, Liu. (2018). Novel microporous β-cyclodextrin polymer as sorbent for solid-phase extraction of bisphenols in water samples and orange juice, *Talanta*, *187*, 207-215.

[22] Somenath, Mitra. (2004). *Sample Preparation Techniques in Analytical Chemistry*, John Wiley & Sons, *162*, 85-86.

[23] Pawliszyn, J. (2002). Sampling and Sample Preparation for Field and Laboratory, *Comprehensive Analytical Chemistry*, *37*, 1031.

[24] Nigel, J. K. Simpson. (2000). *Solid-Phase Extraction: Principles, Techniques, and Applications*, Varian Associates: Harbor City, California.

[25] Steven, H. Y. Wong. & Irving, Sunshine. (1996). *Handbook of Analytical Therapeutic Drug Monitoring and Toxicology*, CRC Press-Taylor & Francis Group.

[26] Núria, Fontanals, Rosa, Maria Marcé. & Francesc, Borrull. (2010). Overview of the novel sorbents available in solid-phase extraction to improve the capacity and selectivity of analytical determinations, *Contributions to Science*, *6* (2), 199–213.

[27] Natalia, Casado, Sonia, Morante-Zarcero, Damián, Pérez-Quintanilla. & Isabel, Sierra. (2018). Evaluation of mesostructured silicas with wormhole-like framework functionalized with hydrophobic groups as alternative sorbents for extraction of drug residues from food samples, *Materials Letters*, *220*, 165–168.

[28] Enric, Pellicer-Castella, Carolina, Belenguer-Sapiña, Pedro, Amorós, Jamal, El Haskouri, José, Manuel Herrero-Martínez & Adela, Mauri-Aucejo. (2018). Study of silica-structured materials as sorbents for organophosphorus pesticides determination in environmental water samples, *Talanta*, *189*, 560-567.

[29] Beshare, Hashemia, Parvin, Zohrabib & Mojtaba, Shamsipur. (2018). Recent developments and applications of different sorbents for SPE and SPME from biological samples, *Talanta*, *187*, 337-347.

[30] Jeffrey, R. Longa & Omar M. Yaghi. (2009). The pervasive chemistry of metal-organic frameworks, *Chemical Society Reviews*, *38*, 1213-1214 DOI: 10.1039/B903811F (Editorial).

[31] Gu, Z. Y., Chen, Y. J., Jiang, J. Q. & Yan, X. P. (2018). Metal–organic frameworks for efficient enrichment of peptides with simultaneous exclusion of proteins from complex biological samples, *Chem. Commun.*, *47*, 4787–4789.

[32] Zhang, S., Yao, W., Fu, D., Zhang, C. & Zhao, H. (2018). Fabrication of magnetic zinc adeninate metal–organic frameworks for the extraction of benzodiazepines from urine and wastewater, *Journal of Separation Science*, Volume *41*, Issue 8, 1864-1870.

[33] Zhirong, Zou, Shanling, Wang. Jia, Jia, Fujian, Xub., Zhou, Long. & Xiandeng, Hou. (2016). Ultrasensitive determination of inorganic arsenic by hydride generation-atomic fluorescence spectrometry using Fe3O4@ZIF-8 nanoparticles for preconcentration, *Micro-chemical Journal*, *124*, 578-583.

[34] Lin, Hao, Xing-Li, Liu, Jun-Tao, Wang, Chun, Wang., Qiu-Hua, Wu. & Zhi, Wang. (2016). *Metal-organic framework derived magnetic nanoporous carbon as an adsorbent for the magnetic solid-phase extraction of chlorophenols from mushroom sample*, *27*, 783-788.

[35] Yuling, Hu, Chaoyong, Song, Jia, Liao., Zelin, Huang. & Gongke, Li. (2013). Water stable metal-organic framework packed microcolumn for online sorptive extraction and direct analysis of naproxen and its metabolite from urine sample, *Journal of Chromatography A*, *1294*, 17-23.

[36] Suling, Zhang, Zhe, Jia. & Weixuan, Yao. (2014). A simple solvothermal process for fabrication of a metal-organic framework with an iron oxide enclosure for the determination of organophosphorus pesticides in biological samples, *Journal of Chromatography A, 1371*, 74-81.

[37] Rosa, Dargahi, Homeira, Ebrahimzadeh, Ali, Akbar Asgharinezhad, Alireza, Hashemzadeh & Mostafa, M. Amini. (2017). Dispersive magnetic solid-phase extraction of phthalate esters from water samples and human plasma based on a nanosorbent composed of MIL-101(Cr) metal–organic framework and magnetite nanoparticles before their determination by GC–MS, *Journal of Separation Science, 41*, 948-957.

[38] Ruiyang, Ma, Lin, Hao, Junmin, Wang, Chun, Wang, Qiuhua, Wu. & Zhi, Wang. (2016). Magnetic porous carbon derived from a metal–organic framework as a magnetic solid-phase extraction adsorbent for the extraction of sex hormones from water and human urine, *Journal of Separation Science, 39*, 3571-3577.

[39] Mina, Asiabi, Ali, Mehdinia. & Ali, Jabbari. (2017). Electrospun biocompatible Chitosan/MIL-101 (Fe) composite nanofibers for solid-phase extraction of Δ9-tetrahydrocannabinol in whole blood samples using Box-Behnken experimental design, *Journal of Chromatography A, 1479*, 71-80.

[40] Nuria, Fontanals, Peter, A. G. Cormack, Rosa, M. Marce. & Francesc, Borrull. (2010). Mixed-mode ion-exchange polymeric sorbents: dual-phase materials that improve selectivity and capacity, *Trends in Analytical Chemistry*, Vol. *29*, No. 7.

[41] Chaonan, Huanga, Yun, Lia, Jiajia, Yangc, Junyu, Penga, Jun, Tand, Yun, Fana, Longxing, Wanga. & Jiping, Chen. (2018). Hyperbranched mixed-mode anion-exchange polymeric sorbent for highly selective extraction of nine acidic non-steroidal anti-inflammatory drugs from human urine, *Talanta, 190*, 15-22.

[42] Yun, Li, Chaonan, Huanga, Jiajia, Yang, Junyu, Penga, Jing, Jina, Huilian, Maa. & Jiping, Chen. (2017). Multifunctionalized mesoporous silica as an efficient reversed-phase/anion exchange

mixed-mode sorbent for solid-phase extraction of four acidic nonsteroidal anti-inflammatory drugs in environmental water samples, *Journal of Chromatography A*, *1527*, 10–17.

[43] Stephan, Moyses. & Anton, Ginzburg. (2016). The chromatography of poly(phenylene ether) on a porous graphitic carbon sorbent, *Journ. of Chrom. A.*, *1468*, 136–142.

[44] Nigel, J. K. Simpson. (2000). *Solid-Phase Extraction: Principles, Techniques, and Applications*, Varian Associates: Harbor City, California.

[45] Marie-Claire, Hennion. (2000). Graphitized carbons for solid-phase extraction, *Journ. of Chrom. A*, *885*, 73–95.

[46] Justyna, Płotka-Wasylka, Natalia, Szczepanska, Miguel, de la Guardia. & Jacek, Namiesnik. (2016). Modern trends in solid phase extraction: New sorbent media, *Trends in Analytical Chemistry*, *77*, 23–43.

[47] Herrero-Latorre, C., Barciela-García, J., García-Martín, S. & Pena-Crecente, R. M. (2018). Graphene and carbon nanotubes as solid phase extraction sorbents for the speciation of chromium: A review, *Analytica Chimica Acta*, *1002*, 1-17.

[48] Neda, Baghban, Ali, Mohammad Haji Shabani. & Shayessteh, Dadfarnia. (2012). Solid phase extraction and flame atomic absorption spectrometric determination of trace amounts of cadmium and lead in water and biological samples using modified TiO_2 nanoparticles, *International Journal of Environmental Analytical Chemistry*, Volume *93*, 1367-1380.

[49] Mazaher, Ahmadi, Tayyebeh, Madrakian. & Abbas, Afkhami. (2016). *Solid phase extraction of amoxicillin using dibenzo-18-crown-6 modified magnetic-multiwalled carbon nanotubes prior to its spectrophotometric determination*, Volume *148*, Pages 122-128.

[50] Beatriz, Fresco-Cala, Óscar, Mompó-Roselló, Ernesto, F. Simó-Alfonso, Soledad, Cárdenas. & José, Manuel Herrero-Martínez. (2018). Carbon nanotube-modified monolithic polymethacrylate pipette tips for (micro)solid-phase extraction of antidepressants from urine samples, *Microchimica Acta*, *127*.

[51] Jennifer, Álvarez Méndez, Julia, Barciela García, Rosa, M. Peña Crecente, Sagrario, García Martín. & Carlos, Herrero Latorre. (2011). A new flow injection preconcentration method based on multiwalled carbon nanotubes for the ETA-AAS determination of Cd in urine, *Talanta*, Volume *85*, Issue 5, 2361-2367

[52] Ayman, A. Gouda. & Sheikha, M. Al Ghannam. (2016). Impregnated multiwalled carbon nanotubes as efficient sorbent for the solid phase extraction of trace amounts of heavy metal ions in food and water samples, *Food Chemistry*, *202*, 409-416.

[53] Davood, Bigdelifam, Mohammad, Mirzaei, Mahdi, Hashemi, Mitra, Amoli-Diva, Omid, Rahmani, Parvin, Zohrabi, Zohreh, Taherimaslak. & Mohammad, Turkjokar. (2014). Sensitive spectrophotometric determination of fluoxetine from urine samples using charge transfer complex formation after solid phase extraction by magnetic multiwalled carbon nanotubes, *Analytical Methods*, Issue 21.

[54] Virginia, Moreno, Eulogio, J. Llorent-Martínez, Mohammed, Zougagh. & Angel, Ríos. (2018). Synthesis of hybrid magnetic carbon nanotubes – C18-modified nano SiO_2 under supercritical carbon dioxide media and their analytical potential forsolid-phase extraction of pesticides, *The Journal of Supercritical Fluids*, *137*, 66–73.

[55] Mateusz, Pęgier, Krzysztof, Kilia. & Krystyna, Pyrzyńska. (2018). Enrichment of scandium by carbon nanotubes in the presence of calcium matrix, *Microchemical Journal*, *137*, 371-375.

[56] Somayeh, Tajik. & Mohammad, Ali Taher. (2011). *A new sorbent of modified MWCNTs for column preconcentration of ultra trace amounts of zinc in biological and water samples*, *278*, Issues 1–3, 57-64.

[57] Hagestam, I. H. & Pinkerton, T. C. (1986). Production of "internal surface reversed-phase" supports: the hydrolysis of selected substrates from silica using chymotrypsin, *J. Chromatogr. A*, *368*, 77.

[58] Petr, Sadílek, Dalibor, Satínsky. & Petr, Solich. (2007). Using restricted-access materials and column switching in high-

performance liquid chromatography for direct analysis of biologically-active compounds in complex matrices, *Trends in Analytical Chemistry*, Vol. *26*, No. 5.

[59] Wayne, M. Mullett. (2007). Determination of drugs in biological fluids by direct injection of samples for liquid-chromatographic analysis, *J. Biochem. Biophys. Methods*, *70*, 263–273.

[60] Souverain, S., Rudaz, S. & Veuthey, J. L. (2004). Restricted access materials and large particle supports for on-line sample preparation: an attractive approach for biological fluids analysis, *Journal of Chromatography B*, *801*, 141-156.

[61] Henrique, Dipe de Faria, Lailah, Cristina de Carvalho Abrao, Mariane, Gonçalves Santos, Adriano, Francisco Barbosa. & Eduardo, Costa Figueiredo. (2017). New advances in restricted access materials for sample preparation: A review, *Analytica Chimica Acta*, *959*, 43-65.

[62] Juan, He, Lixin, Song, Si, Chen, Yuanyuan, Li, Hongliang, Wei, Dongxin, Zhao, Keren, Gu. & Shusheng, Zhang. (2015). Novel restricted access materials combined to molecularly imprinted polymers for selective solid-phase extraction of organophosphorus pesticides from honey, *Food Chemistry*, *187*, 331-337.

[63] Valéria, Maria Pereira Barbosa, Adriano, Francisco Barbosa, Jefferson, Bettini, Pedro, Orival Luccas. & Eduardo, Costa Figueiredo. (2016). Direct extraction of lead (II) from untreated human blood serum using restricted access carbon nanotubes and its determination by atomic absorption spectrometry, *Talanta*, *147*, 478-484.

[64] Fabio, Augusto, Leandro, W. Hantao, Noroska, G. S. Mogollon. & Soraia, C. G. N. Braga. (2013). New materials and trends in sorbents for solid-phase extraction, *Trends in Analytical Chemistry*, Vol. *43*.

[65] Nuria, Gilart, Francesc, Borrull, Nuria, Fontanals. & Rosa, Maria Marce. (2014). Selective materials for solid-phase extraction in environmental analysis, *Trends in Environmental Analytical Chemistry*, *1*, 8–18.

[66] Majid, Soleimania, Serveh, Ghaderia, Majid, Ghahraman Afshara. & Saeed, Soleimani. (2012). Synthesis of molecularly imprinted polymer as a sorbent for solid phase extraction of bovine albumin from whey, milk, urine and serum, *Microchemical Journal, 100*, 1-7.

[67] Jiajia, Yang, Yun, Li, Jincheng, Wang, Xiaoli, Sun, Rong, Cao, Hao, Sun, Chaonan, Huang. & Jiping, Chen. (2015). Molecularly imprinted polymer microspheres prepared by Pickering emulsion polymerization for selective solid-phase extraction of eight bisphenols from human urine samples, *Analytica Chimica Acta, 872*, 35–45.

[68] Ningli, Wu, Zhimin, Luo, Yanhui, Ge, Pengqi, Guo, Kangli, Du, Weili, Tang, Wei, Du, Aiguo, Zeng, Chun, Chang. & Qiang, Fu. (2016). A novel surface molecularly imprinted polymer as the solid-phase extraction adsorbent for the selective determination of ampicillin sodium in milk and blood samples, *Journal of Pharmaceutical Analysis, 6*, 157–164.

[69] Shoulian, Wei, Jianwen, Li, Yong, Liua. & Jinkui, Ma. (2016). Development of magnetic molecularly imprinted polymers with double templates for the rapid and selective determination of amphenicol antibiotics in water, blood, and egg samples, *Journal of Chromatography A, 1473*, 19–27.

[70] Hongliang, He, Xiaoli, Gu, Liying, Shi, Junli, Hong, Hongjuan, Zhang, Yankun, Gao, Shuhu, Du. & Lina, Chen. (2015). Molecularly imprinted polymers based on SBA-15 for selective solid-phase extraction of baicalein from plasma samples, *Analytical and Bioanalytical Chemistry*, Volume *407*, Issue 2, 509–519.

[71] Saman, Azodi-Deilami, Majid, Abdouss, Ebadullah, Asadi, Alireza, Hassani Najafabadi, Sadegh, Sadeghi, Sina, Farzaneh. & Somayeh, Asadi. (2014). Magnetic molecularly imprinted polymer nanoparticles coupled with high performance liquid chromatography for solid-phase extraction of carvedilol in serum samples, *Journal of Applied Polymer Science*, Volume *131*, Issue 23.

[72] Anna, M. Chrzanowska, Anna, Poliwoda. & Piotr, P. Wieczorek. (2015). Surface molecularly imprinted silica for selective solid-phase

extraction of biochanin A, daidzein and genistein from urine samples, *Journal of Chromatography A*, *1392*, 1-9.

[73] Katarína, Hroboňová, Andrea, Machyňáková. & Jozef, Čižmárik. (2018). Determination of dicoumarol in Melilotus officinalis L. by using molecularly imprinted polymer solid-phase extraction coupled with high performance liquid chromatography, *Journal of Chromatography A*, *1539*, 93-102.

[74] Deli, Xiao, Pierre, Dramou, Nanqian, Xiong, Hua, Hea, Hui, Li, Danhua, Yuan. & Hao, Dai. (2013). Development of novel molecularly imprinted magnetic solid-phase extraction materials based on magnetic carbon nanotubes and their application for the determination of gatifloxacin in serum samples coupled with high performance liquid chromatography, *Journal of Chromatography A*, *1273*, 44-53.

[75] Homeira, Ebrahimzadeh, Zahra, Dehghani, Ali, Akbar Asgharinezhad, Nafiseh, Shekari. & Karam, Molaei. (2013). Determination of haloperidol in biological samples using molecular imprinted polymer nanoparticles followed by HPLC-DAD detection, *International Journal of Pharmaceutics*, *453*, 601–609.

[76] Mohammad, Mahdi Moein, Mehran, Javanbakht. & Behrouz, Akbari-adergani. (2014). Molecularly imprinted polymer cartridges coupled on-line with high performance liquid chromatography for simple and rapid analysis of human insulin in plasma and pharmaceutical formulations, *Talanta*, Volume *121*, 30-36.

[77] Silindile, Senamile Zunngu, Lawrence, Mzukisi Madikizela, Luke, Chimuka. & Phumlane, Selby Mdluli. (2017). Synthesis and application of a molecularly imprinted polymer in the solid-phase extraction of ketoprofen from wastewater, *Comptes Rendus Chimie*, Volume *20*, Issue 5, 585-591.

[78] Mohammad, Behbahani, Saman, Bagheri, Mostafa, M. Amini, Hamid, Sadeghi Abandansari, Hamid, Reza Moazami. & Akbar, Bagheri. (2014). Application of a magnetic molecularly imprinted polymer for the selective extraction and trace detection of

lamotrigine in urine and plasma samples, *Journal of Separation Science*, Volume *37*, Issue 13, 1610-1616.

[79] Martinez-Sena, T, Armenta, S, Guardia, M. & Esteve-Turrillas, F. (2016). Determination of non-steroidal anti-inflammatory drugs in water and urine using selective molecular imprinted polymer extraction and liquid chromatography, *J Pharm. Biomed. Anal, 131*, 48-53.

[80] Ge, Chen, Maojun, Jin, Pengfei, Du, Chan, Zhang, Xueyan, Cui, Yudan, Zhang, Yongxin, She, Hua, Shao, Fen, Jin, Shanshan, Wang, Lufei, Zheng. & Jing, Wang. (2017). A sensitive chemiluminescence enzyme immunoassay based on molecularly imprinted polymers solid-phase extraction of parathion, *Analytical Biochemistry*, *530*, 87-93.

[81] Nezhadali, A., Es'haghia, Z. & Khatibi, A. (2016). Selective extraction of progesterone hormones from environmental and biological samples using a polypyrrole molecularly imprinted polymer and determination by gas chromatography, *Analytical Methods*, Issue 8.

[82] Xiaomeng, Su, Xiaoyan, Li, Junjie, Li, Min, Liu, Fuhou, Lei, Xuecai, Tan, Pengfei, Li. & Weiqiang, Luo. (2015). Synthesis and characterization of core–shell magnetic molecularly imprinted polymers for solid-phase extraction and determination of Rhodamine B in food, *Food Chemistry*, *171*, 292-297.

[83] Nageswara Rao, R., Pawan, K. Maurya. & Sara, Khalid. (2011). Development of a molecularly imprinted polymer for selective extraction followed by liquid chromatographic determination of sitagliptin in rat plasma and urine, *Talanta*, *85*, 950-957.

[84] Zhuomin, Zhang, Wei, Tan, Yuling, Hu & Gongke, Li. (2011). Simultaneous determination of trace sterols in complicated biological samples by gas chromatography–mass spectrometry coupled with extraction using β-sitosterol magnetic molecularly imprinted polymer beads, *Journal of Chromatography A*, Volume *1218*, Issue 28, 4275-4283.

[85] Nishide, H., Deguchi, J. & Gladis, J. M. (1976). *Chemical Letters*, *69*.

[86] Bingshan, Zhao, Man, He, Beibei, Chen. & Bin, Hu. (2015). Novel ion imprinted magnetic mesoporous silica for selective magnetic solid phase extraction of trace Cd followed by graphite furnace atomic absorption spectrometry detection, *Spectrochimica Acta Part B*, *107*, 115–124.

[87] Khoddami, N. & Shemirani, F. (2016). A new magnetic ion-imprinted polymer as a highly selective sorbent for determination of cobalt in biological and environmental samples, *Talanta*, *146*, 244–252.

[88] Hamid, Fazelirad, Mohammad, Ali Taher. & Hamid, Ashkenani. (2014). Use of nanoporous Cu(II) Ion Imprinted Polymer as a new sorbent for preconcentration of Cu(II) in water, biological, and agricultural samples and its determination by electrothermal atomic absorption spectrometry, *Journal of AOAC International*, Volume *97*, 4, 1159-1166.

[89] Zhong, Zhang, Jinhua, Li, Xingliang, Song, Jiping, Mac. & Lingxin, Chen. (2014). Hg^{2+} ion-imprinted polymers sorbents based on dithizone–Hg^{2+} chelation for mercury speciation analysis in environmental and biological samples, *RSC Advances*, Issue 87, 46444-46453.

[90] Tarley, C. R. T., Andrade, F. N., De Oliveira, F. M., Corazza, M. Z., De Azevedo, L. F. M. & Segatelli, M. G. (2011). Synthesis and application of imprinted polyvinylimidazole-silica hybrid copolymer for Pb^{2+} determination by flow-injection thermospray flame furnace atomic absorption spectrometry, *Analytica Chimica Acta*, *703*, 145–151.

[91] Kiani, A. & Ghorbani, M. (2017). Synthesis of core–shell magnetic ion-imprinted polymer nanospheres for selective solid-phase extraction of Pb2+ from biological, food, and wastewater samples, *J. Dispers. Sci. Technol.*, *38*, 1041–1048.

[92] Farid, Shakerian, Shayessteh, Dadfarnia. & Ali, Mohammad Haji Shabani. (2012). Synthesis and application of nano-pore size ion

imprinted polymer for solid-phase extraction and determination of zinc in different matrices, *Food Chemistry, 134*, 488-493.

[93] Luz, E. Vera-Avila, Laura, Rangel-Ordoñez. & Rosario, Covarrubias-Herrera. (2003). Evaluation and Characterization of a Commercial Immunosorbent Cartridge for the Solid-Phase Extraction of Phenylureas from Aqueous Matrices, *Journal of Chromatographic Science, 41*, 480-488.

[94] Maud, Bonichona, Audrey, Combès, Charlotte, Desoubries, Anne, Bossée. & Valérie, Pichon. (2017). Development of immunosorbents coupled on-line to immobilized pepsin reactor and micro liquid chromatography–tandem mass spectrometry for analysis of butyrylcholinesterase in human plasma, *Journal of Chromatography A, 1526*, 70–81.

INDEX

HIGH-PERFORMANCE LIQUID CHROMATOGRAPHY: TYPES, PARAMETERS AND APPLICATIONS

EDITOR: Ivan Lucero

SERIES: Analytical Chemistry and Microchemistry

BOOK DESCRIPTION: In this collection, the authors discuss the way in which it is possible to detect the low concentrations at which toxic compounds and metabolites are present in specimens due to the huge development of chromatographic techniques.

SOFTCOVER ISBN: 978-1-53613-543-5
RETAIL PRICE: $95

HYDRIDES: TYPES, BONDS AND APPLICATIONS

EDITOR: Patrick C. Dam

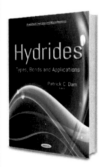

SERIES: Analytical Chemistry and Microchemistry

BOOK DESCRIPTION: *Hydrides: Types, Bonds and Applications* first proposes metal hydrides as a fascinating class of compounds due to the small mass and size of hydrogen. Its medium electronegativity causes a large flexibility in terms of metal-ligand interactions, resulting in a vast variety of possible compositions, chemical bonding, crystal structures and physical properties.

HARDCOVER ISBN: 978-1-53613-581-7
RETAIL PRICE: $160

X-Ray Fluorescence: Technology, Performance and Applications

EDITOR: Rebeca Fonseca

SERIES: Analytical Chemistry and Microchemistry

BOOK DESCRIPTION: *X-Ray Fluorescence: Technology, Performance and Applications* opens with a study wherein the possibilities of polarizing energy dispersive X-ray fluorescence spectrometry were fully explored to materialize rapid trace element determinations of various geological samples.

SOFTCOVER ISBN: 978-1-53614-303-4
RETAIL PRICE: $82

Flow and Capillary Electrophoretic Analysis

EDITORS: Paweł Kościelniak (Jagiellonian University in Krakow, Krakow, Poland); Marek Trojanowicz (Institute of Nuclear Chemistry and Technology, and University of Warsaw, Warsaw, Poland)

SERIES: Analytical Chemistry and Microchemistry

BOOK DESCRIPTION: This book presents current development trends in flow analysis and capillary electrophoresis. It contains numerous review chapters dedicated to various aspects of both techniques.

HARDCOVER ISBN: 978-1-53613-184-0
RETAIL PRICE: $230